# GM Agriculture and Food Security

## Fears and Facts

# GM Agriculture and Food Security

## Fears and Facts


**Stuart J. Smyth**

**William A. Kerr**

and

**Peter W.B. Phillips**


CABI

**CABI is a trading name of CAB International**

CABI
Nosworthy Way
Wallingford
Oxfordshire OX10 8DE
UK

Tel: +44 (0)1491 832111
Fax: +44 (0)1491 833508
E-mail: info@cabi.org
Website: www.cabi.org

CABI
745 Atlantic Avenue
8th Floor
Boston, MA 02111
USA

T: +1 (617)682-9015
E-mail: cabi-nao@cabi.org



A catalogue record for this book is available from the British Library, London, UK.

**Library of Congress Cataloging-in-Publication Data**

Names: Smyth, Stuart, author. | Kerr, William A. (William Alexander), author. | Phillips, Peter W. B., author. | C.A.B. International, publisher.
Title: GM agriculture and food security : fears and facts / Stuart J. Smyth, William A. Kerr and Peter W.B. Phillips.
Description: Oxfordshire, UK ; Boston, MA : CAB International, [2019] | Includes bibliographical references and index.
Identifiers: LCCN 2019021317| ISBN 9781786392213 (hardback) | ISBN 9781786392244 (paperback) | ISBN 9781786392237 (epub) | ISBN 9781786392220 (pdf)
Subjects: | MESH: Food Supply | Food Technology | Food, Genetically Modified | Food Safety | Food Labeling
Classification: LCC HD9005 | NLM WA 695 | DDC 338.1/9--dc23 LC record available at https://protect-eu.mimecast.com/s/ATPCCl2gvHoAx4IGpQFB?domain=lccn.loc.gov

ISBN-13: 978 1 78639 221 3 (hardback)
978 1 78639 224 4 (paperback)
978 1 78639 222 0 (ePDF)
978 1 78639 223 7 (ePub)

Commissioning Editor: David Hemming
Editorial Assistant: Emma McCann
Production Editor: Kate Hill

Typeset by Exeter Premedia Services Pvt Ltd, Chennai, India
Printed and bound in the UK by Bell & Bain Ltd, Glasgow

# Contents

# Abbreviations

| | |
|---|---|
| ABS | access and benefits sharing |
| ALA | α-linolenic acid |
| Bt | *Bacillus thuringiensis* |
| CBD | Convention on Biological Diversity |
| CEC | Commission for Environmental Cooperation |
| CFIA | Canadian Food Inspection Agency |
| CGIAR | Consultative Group for International Agricultural Research |
| CJEU | Court of Justice of the European Union |
| COOL | country of origin labelling |
| CUDOS | Communalism, universalism, disinterestedness and organized scepticism |
| DDT | dichlorodiphenyltrichloroethane |
| DHA | docosahexaenoic acid |
| DNA | deoxyribonnucleic acid |
| DSI | digital sequencing information |
| EFSA | European Food Safety Authority |
| eNGO | environmental non-governmental organization |
| EPA | eicosapentaenoic acid |
| EU | European Union |
| FAO | United Nations Food and Agriculture Organization |
| GATT | General Agreement on Tariffs and Trade |
| GCDT | Global Crop Diversity Trust |
| GM | genetically modified |
| GMO | genetically modified organism |
| GODAN | Global Open Data for Agriculture and Nutrition |
| HRAC | Herbicide-Resistance Action Committee |
| HT | herbicide tolerant |
| IARC | International Agency for Research on Cancer |
| ICABR | International Consortium of Applied Bioeconomy Research |
| IPRs | intellectual property rights |
| IR | insect resistant |

| | |
|---|---|
| MNE | Multi national enterprise |
| MRL | maximum residue limit |
| NAFTA | North American Free Trade Agreement |
| NGO | non-governmental organization |
| NOEL | no-observable-effect level |
| OECD | Organisation for Economic Cooperation and Development |
| PBR | plant breeders' rights |
| PGR | plant genetic resource |
| PMRA | Pest Management Review Agency |
| PNT | plant with novel traits |
| QALY | quality-adjusted life year |
| R&D | research and development |
| RNA | ribonucleic acid |
| SPS | Application of Sanitary and Phytosanitary Measures Agreement |
| TB | tuberculosis |
| TRIPS | trade-related aspects of intellectual property rights |
| UN | United Nations |
| UPOV | International Union for the Protection of New Varieties of Plants |
| USDA | United States Department of Agriculture |
| USRTK | United States Right to Know group |
| VR | virus resistant |
| WHO | World Health Organization |
| WTO | World Trade Organization |

# Preface

We three authors have worked together for nearly 20 years, drawing on extensive backgrounds in academic research, the agri-food industry and government. We also live in Saskatchewan, Canada, a major agri-food producing and exporting region of the world. Saskatoon, the city in which we live and work, brought the first genetically modified canola and flax varieties to the global market and has been a testbed for many of the new biosciences over the past half century. As a result, we have lived at the frontier of the controversies about the future of the global food system, since the products of modern biotechnology were under development and first brought to the market on a commercial basis.

In the past, as scholars, we have tried both individually and collectively to bring more structured logic and empirical evidence to the debate about where and who in the broadly defined 'global food supply chain' should be able to use the bundle of new biosciences, and how they should be used. While we think our scholarly work has had some modest effect in some quarters, overall, we see a world that is too often following rumours, innuendoes and trails of false evidence.

When we discussed the purpose of this book, we came to the realization that any evidence presented, no matter how refined, theoretically grounded, methodologically rigorous or empirically informed, was unlikely to have much impact. When we discussed our ideas with a colleague, she responded that laypersons are 'concerned about the buzz words of *GMO*, *organic*, *natural*, *environment*, and *experts*, that they hear on a regular basis, but fail to connect the dots on how it directly affects them'. She put the challenge into stark relief: 'As they munch on their gluten-free avocado toast and quinoa salad, accompanied by their daily latte, they declare that shopping local is the best thing ever. They have no idea that none of what they just ate can realistically be grown here. There is a nearly complete disconnect between consumers and the agri-food supply chain.' This volume attempts to reconnect the dots. To do that, we have set aside our deductive theories, complex methods and reams of data and have tried to explain our perspective in plain English.

We hope you find new insight and meaning in our work. We enjoyed the new-found freedom of simply writing down what we understand. Most of what we say can be backed up by our research and the work of the larger policy community. A set of further readings are offered at the end of each chapter to assist those who want to dig deeper into any topic. In a few cases however, we have found missing links in the logic – in those cases, we have used our more than 100 combined years of experience to fill in the gaps.

We would like to thank all the anonymous reviewers of our work, the global network of scholars who present and debate at the International Consortium of Applied Bioeconomy Research (ICABR), and our friends, colleagues and family who were the first audience for many of our arguments in this book. We would especially like to thank Sara Alexander, Wanda Phillips, Lori Sheremata and May Yeung for reading the final product and offering their advice. As always, any errors of commission or omission are ours and ours alone.

**SJS, WAK and PWBP**

# 1 Food Security and Technology: Fear Trumps Hope

*As we sat down to sketch out the outline for this book in 2017, we thought that the best way to add to the debate about the role of biotechnology in the world was to present our views, grounded in our theoretical understandings and our research results, but translated through our personal experiences. We were motivated to present our understanding of the evidence in ways that would reach new audiences. We started from the perspective that a vigorous scientific defence would be the best approach. In essence, we aspired to present a work of 'rationalistic aggressiveness'. We sincerely believed that evidence speaks louder than all other arguments, especially those based on half-truths and untruths.*

*At root, this approach assumes that rationality is the underlying metaphor for discourse, deliberation and decision making in the 21st century. As we headed into our respective writing spaces, the world we hoped to influence changed. The election of Donald J. Trump as president of the USA in November 2016 and his first 2 years in office have signalled a new style of political discourse. In this new era, few start their arguments with theory or evidence. World views, beliefs, aspirations and interests define the narrative, and evidence, if used at all, now simply supplements and supports rather than defines the discourse. Facts are negotiable, or at least open to selection. Admittedly, aggressively propounded, self-interested and biased arguments have been the hallmark of politics since governance was opened to the many (i.e. democracy). With a few historical exceptions, the give and take in political debate, an active and critical media and a naturally sceptical and balanced citizenry and electorate sorted the wheat from the chaff. Rhetorical or self-interested excesses were winnowed out. The net result was a perception that we had moved to a higher standard of discourse and decision making in the public space and in corporate headquarters around the world.*

*Mr Trump's behaviour since 2016 challenges this understanding of the world. In many ways, we are in a world reminiscent of the fictional Oceania in George Orwell's Nineteen Eighty-Four, where the 'Party', in the name of its leader, 'Big Brother', employed 'Thought Police' to persecute individualism and independent thinking. 'Doublethink', 'thoughtcrimes', 'Newspeak' and 'memory holes' offered new terminology to describe a world of official deception, brazenly misleading language and manipulation of recorded history by a totalitarian leader. Although few nation states have achieved enough power to control our thoughts in such a deceptive way, many have joined opposing teams and act as if they have bought into that approach. This self-constructed groupthink is twisting the policy landscape in ways that make evidence-based discourse and policy making profoundly more challenging.*

*This introductory chapter explores the clash between the rhetorical flourishes and the objective evidence that frames the global policy dialogue directed at the challenge of securing an adequate and sustainable supply of affordable and nutritious food for a growing world population.*

## Our Working Hypothesis

While we will not present our discussions in an academic way, we will draw on the literature from public policy and agenda setting, in which theorists and practitioners have posited that ill-structured problems are publicly debated and defined in an increasingly diverse world of actors who often remotely engage through the Internet in an effort to promote their value-laden causal stories. Decision making is then pursued in ways that deliver outcomes that can work at cross purposes to our stated public objectives.

Our goal is to use this book to reconfigure the underlying architecture of public consideration of science, technology and commerce in the global food system. We will explore an array of characterizations of the space, ranging from Frankenfoods and superweeds to superfoods and miracle plants. Ultimately, we want to open the discussion about how we can construct a factual dialogue about complex issues that uses evidence to help make better choices.

## The 'Rule of Three' as a Motivator

In many ways, the rule of three underlies our work. Mythology, religion, literature, art and the law all posit that three elements are more likely to deliver better and more satisfying outcomes. This is formalized as a writing principle, whereby a triad of events or characters is more effective, engaging and enduring than other configurations. To this end, we are three authors, who collectively span three different decades of life and three complementary perspectives on the world.

Economist Kenneth Boulding asserted that humans live in three somewhat differentially motivated worlds. We are, simultaneously, citizens, economic beings and social animals. Citizenship in a country is for most of us compulsory, and the state is the only actor who can legitimately coerce citizens to comply with state direction. Citizenship motivates our common interests. Our economic selves engage in voluntary exchange through markets – theorists argue that in a perfect world markets won't need any external guidance to deliver mutually beneficial transactions. As social beings we belong to groups that define our identity, meaning and purpose, such as school friends, our workplace and cultural or recreational groups.

Each of us as authors and scholars has a different balance of passions, experiences and training. We all have the numeraire of training as economists and employment in the academic profession, but Bill has been an academic economist for 40 years, contributing to the body of evidence and the policy space through teaching, research and active engagement. Stuart, our youngest, had a first career as a farmer and business consultant, before pursuing his doctorate and joining the professoriate. Peter started as a professional economist in industry, moved to executive government and then in mid-career joined the academy, with appointments in agricultural economics, economics, policy, political science and sociology. Each of us appreciates the multidimensional space but has a core interest. Bill focuses on the market and exchanges, and explores how the other domains can support or frustrate markets doing what they do best; Peter is concerned about legitimate and appropriate governance and how individuals, through the state, markets and collective action, make choices; and Stuart is differentially engaged with the diffuse and difficult conversations that bubble up from society and push the state and market away from their core goals.

While we have different orientations, we have a few common commitments. We are unapologetic experts, in that we believe objective evidence can and should be the foundation, if not the driver, of collective decisions. Moreover, we each respect the role of Boulding's three domains. States are the only legitimate and capable actors to deal with public goods and to adjudicate many disputes, markets are uniquely positioned to efficiently allocate scarce resources among competing ends and the social space, where we create norms, is necessary for getting almost everything done.

## Our Modernist Foundation

We are unapologetic modernists. We firmly believe that some things can be known objectively, and that opinions, interests or beliefs will not change that reality. Gravity is the classical objective reality – as we stand on the edge of a drop, we instinctively know that gravity and its potentially fatal effects are not just beliefs, but real. We know that this is not fashionable; the modernist search for facts, laws of nature and empirically consistent cause and effect relationships, especially in the economic and social space, is hotly contested by many. Nevertheless, we will not apologize, but rather explain what modernism is, why we have faith in it and why we offer it as a way forward to a better future. We are not, however, slaves to the conventional view of modernism as uniquely motivated by what political scientist Deborah Stone called the 'rationality project'. We see the search for objective evidence as a necessary but far from sufficient condition for progress in our economic, social, political and environmental entanglements.

Isaac Newton wrote in 1676, 'If I have seen farther it is by standing on the shoulders of giants.' In his day, finding the giants on whose shoulders to stand must have been somewhat easier, as there were only a few sites of higher learning and a relatively small number of thinkers, scholars and inventors. With the advent of the earned doctorate in the 19th century, and the sharp escalation in university enrolment and public and private research in the past century, there are now millions of scholars and practitioners who may be the source of new knowledge and ideas. This expansion and diffusion of effort makes it difficult (some assert impossible) to catalogue, assess, integrate and disseminate the mass of knowledge that we know or might want to know.

There are many actors involved in governing the normalization of emerging knowledge in the natural sciences, applied technologies and social sciences, making it increasingly difficult to establish what is known, what is new, where it is going and how it might be applied safely and efficaciously. Political scientist Sheila Jasanoff suggests that knowledge is ultimately 'co-produced' by many constantly intertwined actors at four stages: emergence, contestation, standardization and enculturation. Two main types of knowledge provide the foundation for what we know: codified knowledge, such as the 'know-why' knowledge about the physical and social world around us (e.g. the laws of nature embodied in academic treatises); and tacit, 'know-how' knowledge about the way to do something (often embodied in individual scholars and practitioners). Both types of knowledge involve codifying and preserving a shared set of normative and principled axioms, shared causal beliefs, notions of weighing and validating knowledge in each domain and a common set of practices associated with a specific set of problems. This process is embodied in both the conceptual framing of 'normal science' and the practice of individual governing bodies in the academies and professions. The process of assessing and weighing new knowledge is challenging. In many cases the response to new ideas will be purpose-built processes to address the new knowledge claims.

Over millennia, society has developed a series of governing systems (some might call them subroutines) to undertake the task of codifying, assessing, integrating and disseminating our knowledge stock. This system has involved a range of disciplines, tools, systems and institutions, such as taxonomies (e.g. the Linnaean classification of species, the periodic table of elements, disciplinary codes used by granting agencies and journals and the Dewey decimal system in libraries), languages (e.g. mathematics and the ACGT codes of genes), and institutions (e.g. universities, public research organizations, journals and peer review).

In the past, new scientific developments were tried, tested and challenged in a slow, methodical way outside the limelight. The underlying principle of science is that the results of one experiment must be repeatable by others, which requires time to ensure that the results of any particular experiment actually deliver credible, meaningful, repeatable results. The goal of science, however, is no longer simply discovery and dissemination of new knowledge; increasingly scientists want broader recognition and commercial gain. More than half of research in many domains is now undertaken either by private-sector scientists or by public-sector and university scientists funded by, and working collaboratively with, private interests. As an example, our university, the University of Saskatchewan, has 118 research chairs, 36 funded by sources external to the university. Now, each new discovery (touting either new unimaginable benefits or unfathomable risks) is put in front of a largely scientifically unsophisticated public with great fanfare in glowing press releases through traditional media outlets and on the Internet, hyped lead articles in learned journals, perfunctory front-page stories in tabloids and quality broadsheets, and brief, sensationalized sound bites on radio and television, not to mention the plethora of inaccurate, fear-based social media postings. Tim Caulfield, a legal ethicist at the University of Alberta, has described this process in the biotechnology area as 'genohype'. A circus-like atmosphere characterized by excitement and fear has replaced the traditional period of contemplation, assessment, testing and re-testing, complicating our understanding of new technology.

## Assessing Modernist Knowledge Claims

Thomas Kuhn, in *The Structure of Scientific Revolutions*, laid out a useful framework for thinking about the structure for existing knowledge and the staging of new knowledge. He provides a generalized picture of the process by which new knowledge is born, undergoes development and eventually becomes accepted as fact by other scientists and eventually the public. Scientific development according to Kuhn's model involves the study of phenomena, which leads scientists to develop explanatory theories that, after an appropriate period of challenge and testing, lead to a 'mature' science and, if sustained through subsequent challenges, become paradigms. In essence, a paradigm is a set of assumptions about how things work; some might call this a world-view. As a matter of course, the evolutionary process involves a period of 'normal science', when efforts are made to articulate the paradigm, explore possibilities within the paradigm, use the theory to predict facts, solve scientific puzzles, and develop new applications of theory. Throughout this period, science is challenged to address both known phenomena and unknown possibilities. This period of new evidence and expanding application of theory all takes place within the paradigm of normal science.

One way to frame this is to recall how you may have interacted with the Rubik's cube puzzle. Many of us were perpetually frustrated by something that was 'obviously' utterly unsolvable. Eventually, after considerable trial and error, we might solve one side of the cube. This was an individual triumph. Through repeated efforts, we learned how to routinely solve the cube so that one colour could be placed together. Once we determined how to solve the cube with one colour, we could progress to solving it with two or more colours. Some people were able to identify a pattern to the process and ultimately solved the cube such that all six sides were properly aligned. Others showed innovative strategies, such as peeling all the stickers off and reattaching them (the guilty will remain anonymous).

The state of normal science is challenged when new phenomena are observed. For the most part, existing theory and methodology usually can adequately and successfully explain these observations. At times, however, there may be further discoveries of natural phenomena that violate the expectations that govern normal science. This involves discovery of problems not previously known to have existed. In some cases, normal science cannot fully explain them. A relatively abrupt shift or transition to a new paradigm would then become necessary, which would require someone to define and exemplify a new conceptual and methodological framework.

Although Kuhn's examples of the formation and transformation of paradigms were drawn entirely from the history of the physical sciences, his model can be applied to almost any discipline that seeks to frame itself as science. Both theory and evidence, fundamental features of any science, can be known or unknown. Thus, a classic quadrant box can be used to describe the relationships between known and unknown theory and evidence, highlighting four possible realms of scientific investigation, which have different membership and control mechanisms that can vary in importance depending on the science involved: conventional/normal, experimental, hypothesis-driven or speculative.

*Conventional* or *normal* science operates where there is a comprehensive paradigm (with both theory and methodology) that offers predictions that are borne out in the evidence. At the most basic level of this, theory exists stating that $1 + 1 = 2$, there is a formally recognized method (mathematics) that is capable of testing this theory, and ultimately there is proof that when 1 and 1 are added together, they do indeed result in 2. In these cases, it is possible that new evidence or enhancement of the theories or methodologies could offer finer conclusions, but the general conclusions remain constant. This science is almost exclusively the preserve of the global academic community, having been developed, stored and transmitted from universities and other peer-reviewed research centres. The internal control systems of the global scientific community – namely the peer-reviewed structure of appointment, tenure, promotion and publication – enforce rigour and conformity, reducing the chance for unsubstantiated theory or evidence being perpetuated over long periods. During periods of stability, this science dominates.

*Experimental* science offers the most frequent challenge to conventional or normal science. As detection systems become refined (e.g. new genomic, proteomic and metabolomic methodologies in the food system), there is an increasing likelihood that new evidence may be found that is not in the first instance consistent with conventional theory. This evidence could either challenge theory directly by the discovery of the presence (or absence) of some element that theory says should not (or should) have been found, or indirectly, by illuminating some impact that is not formally incorporated into the theory. For decades, it was thought that there was no way to detect planets around distant stars. With advanced telescopes (some in space), we were able to detect 'wobbles' in the light emitted from stars. Planetary scientists theorized that these 'wobbles' were caused by planets transecting the stars. This stage of science has historically been very important, as researchers have sought to discover phenomena that need explaining. The challenge is that while some of this experimental work is done in the academic world, much of it by necessity is done by practitioners or laypersons working in the market and society. These practitioners tend to be isolated from the global research community, as they are examining *in situ* events which may only have local or regional impact. The difficulty with this is that the internal governance/control mechanisms are often relatively weak. Measurement and detection methodologies vary widely depending on what they are designed for. The evidence is not always fully disclosed and quite often there is no formal peer-review structure to assess the significance of new experimental results, either in statistical or scientific terms. This raises the potential of erroneous measurement of irrelevant evidence that can be disseminated through industry or the press or through less formal channels, such as the newsletters of non-governmental organizations (NGOs) or the Internet.

Challenges also come from the realm of *hypothesis-driven* science. As new knowledge and techniques accumulate, they raise new questions about the assumptions and structures of normal or conventional science. The challenges range from fundamental to incremental. New hypotheses often emerge when different parts of normal science are combined to address issues where conventional science may be silent. The underlying difficulty is that one can never prove any hypothesis or theory – one can only gather evidence that disproves a hypothesis. Absence of disproving evidence helps to support theories, but they remain contestable. While most influential hypothetical science is generated within the academic world with its built-in system for evaluating relevance, two recent trends have complicated that. In the first instance, many fields of science have become part of 'big science', which involves large outlays for machinery, equipment and interdisciplinary teams of scientists. This has made it more difficult for challenges to surface and be tested solely within any discipline or sometimes within the academic community itself. The price of admission to big science is the ability to pair theory with empirical methodologies, which frequently means working with conventional science. There are few funders – public or private – that will make substantial investments in highly uncertain science. Meanwhile, there has been a rise in the number of scientists with advanced degrees working outside academia (in industry, government and NGOs), which, paired with increased competition by the media and some of the more commercial scientific journals, often leads to premature public release of incomplete or often readily refuted hypotheses. These scientists may have an interest in engaging in the scientific process but are not as bound by the rigours of the scientific method as those inside academia. Hence, we often see the promotion of hypotheses regarding benefits or risks of new technologies, that upon further consideration and experimentation are rejected. While new hypotheses are an important part of challenging our understandings, they must be treated with caution as there are probably an infinite number of possibilities to consider.

Finally, there is the realm of *speculative* science, where there is neither well-articulated theory nor any body of evidence to suggest that a causal relationship is present. This subset of science offers great promise but also poses tremendous challenges. It is probably fair to say that many of the great breakthroughs in science have come from speculation that is not linearly tied to either any specific evidence, or to the conventional science of the day. While this science is important, it most certainly needs to be filtered through further experimentation and theorizing before it can provide a base for action. The difficulty is that speculative science is perhaps the most exciting and engaging of all the realms. Hence, it generates significant public scrutiny and debate, often at times when it needs to be considered more fully in the context of the scientific method. The problem frequently experienced in today's instant media world is that when a new idea emerges, people hear about it and take the results as verified. Fears of harm instinctively trump positive new benefits. The regimes for governing this process are frequently undefined initially; often the structures need to be purpose-built to handle the process. Academics engaging in speculative science remain bound by the strictures of peer review but because speculation often arises from those beyond the bounds of any structured community, it can be hard to find a suitable forum for consideration. Equally important, many marginally connected academics and non-academic scientists and commentators have nothing to lose by speculating – in the parlance of the day, any media coverage is good, as long as they spell your name right.

To draw on a sports metaphor, normal scientific knowledge is governed by playing both the individual player and the ball. Individuals who claim knowledge and expertise are for the most part accredited through an educational or training system (such as a university, technical college, professional body, craft union or guild) while the information, knowledge and ideas generated by these people are similarly vetted through a variety of evaluation processes.

Scientific norms are founded on a particular framing of the scientific process that is derived from the past four centuries of practice. Historians Stephen Shapin and Simon Schaffer traced the current structure of science back to its roots in the 1660s, when Robert Boyle, the first president of the Royal Society, laid out the style of experimentation we use today. Boyle proposed that science should be the outcome of a process of engaging in an empirical experiment, which would necessarily include repeated witnessing. The focus on the social aspects of science was an effort to remove science from the realm of alchemy, where individuals claimed great accomplishments but could neither replicate nor explain what happened. In this approach, witnessing included the repeated undertaking of experiments and the experimental essay. The essay, in a plain, ascetic, unadorned (sometimes called functional) style, was to provide adequate information to allow replication and to lay out the specific, narrow findings and implications of the experiment (also including experimental failures). Disputes in this system would then be about findings and not persons. In the end, Boyle's model dominates in the physical and natural sciences. Those who decline to play by the rules were not, and are not today, viewed as legitimately part of science.

Michael Polanyi, a scholar of the philosophy of science who drew on his own experience as a medical doctor, physical chemist and economist, argued that science is organized much like a republic, where authority is essentially mutual, established between scientists. Provided standards of professional merit are adhered to, the pursuit by independent, self-coordinated individuals will be the most efficient organization of science (this is analogous to the economic framing of society, in which self-interested, optimizing individuals are expected to generate socially desirable outcomes). Polanyi offered three standards of professional merit. First, work must be plausible, such that the structuring or results of an experiment must bear some correlation to established knowledge; outliers, while perhaps correct, will be treated as anomalous. Subsequent research is needed to test for methodological flaws in outlier results, thereby either invalidating them or addressing flaws in established knowledge. Second, scientific value is assessed by a balance of accuracy, systemic importance and the intrinsic interest of a subject matter. Third, work should exhibit originality. The balance of the three merits keeps science functioning; plausibility and value tend to be conservative traits whereas originality keeps science current and progressing.

Sociologist Robert Merton argued that four institutional imperatives drive modern science – communalism, universalism, disinterestedness and organized scepticism – which he abbreviated into CUDOS. Science should be commonly owned (communal), with the findings assigned to the community, with any single scientist's claim to intellectual property limited to recognition and esteem. Science is universal, in that scientific claims should be international, impersonal and for all intents anonymous. Good science should also be disinterested, in that the focus is on the verifiability of results rather than the authority of the proponent. Finally, science should be tentative and open to challenge. Scepticism should rule over all else. In a world where science adheres to the CUDOS norms, the free competition of ideas would trump special interests.

Philosopher Karl Popper tried to simplify the challenge, asserting that what distinguishes science from what he terms 'non-science' is whether a discipline uses inductive or deductive reasoning. He asserts that science cannot rely on induction, because it is possible to find evidence to support almost any assertion – humans are hard wired for the so-called confirmatory biases, where we seek information to support our predetermined beliefs. Instead, falsifiability is the mark of true scientific investigation. Every genuine test of a scientific theory, then, would logically be an attempt to refute or to falsify it and one single, genuine falsification would lead one to reject an entire theory.

## Governing 'Knowledge Wars'

Economic historian Joel Mokyr asserted that institutions play a central role in determining which knowledge is sustained. Science uses a set of norms – open science, credit to 'first to discover', reproducibility of experiments, scepticism and witnessing – and a range of institutions that are engaged, in the first instance, to define what we know (and by definition, prescribe what is truly novel) and secondly to create bridges between different knowledge domains. Five main interlocking and overlapping structures have evolved to handle the task. At the most basic level, groups of scholars have created and expanded taxonomies and coding systems to provide a base for storing, accessing and managing information from differing fields. Second, a number of foundational languages have evolved and expanded – mathematics is the universal base for many fields. Third, the educational system that has emerged over the past millennium is a major agent in both controlling and promoting the conservation of existing knowledge and the development and dissemination of new knowledge. Fourth, academic and professional associations have had a major influence in defining the nature of disciplines and the conservation, accumulation and dissemination of knowledge. Finally, the institution of peer review is embedded in most of the other institutions, as a rule of engagement and a unifying principle of knowledge development and preservation.

Sorting and resolving knowledge claims has become more complicated. The challenge here is that there has been a 'democratization' of many other fields, not the least of which is information-sharing. Yet science somehow remains impervious to this and, therefore, appears to not be trustworthy, or at least to be worthy of suspicion. While peer review in university settings used to slowly and methodically sort and affirm the wheat from the chaff in earlier times, now collective organizations that give voice to the myriad views of society have greater input. In absence of clear signals, citizens or consumers, states and markets seek ways to mobilize 'voice' in decision making. Collective organizations are the natural microphones for amplifying and transmitting voice – especially not-for-profit social action groups, which represent segments of society that are disassociated from the republic of science and cannot be adequately represented in the homogenizing processes offered by the state and market. While collective organizations are often very effective in mobilizing voice, they frequently complicate the governing process, which in some instances is their only rationale for existing. Challenges arise when competing civil authorities articulate values, interests and beliefs that are incompatible or difficult to pursue jointly. In those situations, some, if not all, civil authorities are bound to be disappointed and frustrated because the nature of their claim is not subject to compromise.

Transformative innovation is fundamentally about the introduction of new, revolutionary science that changes the way we see our circumstances and our opportunities. As with other change, there are always vested interests, values and circumstances that will lead to reactions against new ideas. In some cases, the forces of conservatism are justified, in that the new candidate knowledge either is incorrectly formulated or simply does not generate any new value to the economy or society. To generate optimal value, it needs to be normalized into a paradigm. However, unlike normal knowledge, for which there are longstanding and well-developed institutions that either individually or collectively delimit the nature of our knowledge, contested knowledge does not have any well-established set of institutions. Rather, an array of ad hoc, often purpose built, approaches are used.

Every modern transformative technological revolution has led to a period of contested knowledge, during which some knowledge claims are readily accepted, some are rejected out of hand and some live on in the hearts and minds of men and women, generating heated debate and confusion about what truth to believe. The area of agricultural biotechnology offers a range of global case studies of how new, dramatic (and some would say antithetical) knowledge claims were introduced, adjudicated and at least partially resolved. When biotechnology entered the agricultural and food system in 1994, expert and public opinion ranged from enthusiastic to indifferent. That situation continued until 1998, when several events shook the foundations of the rapidly expanding global agricultural biotechnology industry. In 1998, a study in the UK raised concerns about the safety of genetically modified (GM) insecticidal potatoes and in 1999 a study in the USA suggested that *Bacillus thuringiensis* (Bt) maize could have unintended adverse effects on the habitat of monarch butterflies. The one-two punch of the two studies unleashed a firestorm of claims, counterclaims and processes to mediate the contested claims. Then in 2001, Quist and Chappela from University of California, Berkeley, released a study claiming that GM traits had been transferred into landrace maize and teosinte in Mexico, which triggered an international challenge to the technology in the context of biosafety and traditional knowledge. Finally, the debate in the farm community was personified by Percy Schmeiser, a Saskatchewan canola farmer, who was sued and lost a case of patent infringement to Monsanto, the leading industrial proponent and developer of GM crops and the owner of the patented Roundup Ready™ technology at the core of the dispute. Each of the events involved objective evidence that largely conformed to 'normal' science but was contested by new actors with new, frequently non-scientific, motives.

In retrospect, the controversies offer a number of lessons about governing contested knowledge. From the beginning, most of these contests exhibited post-modern attributes: each involved the release and debate of scientific and social aspects in ways that exceeded the norms of the scientific method, often involving extraordinary campaigns to influence the outcomes, frequently by referencing authority rather than repeated experimentation or peer evaluation. Each generated lively debate about the nature of science and its use but failed to come up with a clear process for judging knowledge claims in the area of contested science. Scientists themselves challenged the norms – when confronted with perspectives beyond their scientific expertise they tended to drop their normal scientific decorum and present a *mélange* of both normal science and their musings about potential concerns (i.e. their next research issue), without distinguishing between them. Hence, it was seldom clear what was normal science and what was emerging (i.e. hypothetical, experimental or speculative) science. The tenor of the 'post-normal' debate was disturbing. In a direct throwback to earlier times, the debate became personalized; instead of playing the ball (by criticizing the science), many participants played the ball carrier (criticizing the authors or actors in the controversy). Finally, even when the primary issue at hand was resolved, there remained enough doubt and controversy that each event lives today in the global debate about biotechnology and its fit with the rest of the world. This highly charged personification of scientific debate, while perhaps understandable given the perceived interests at stake, does little to clarify the claims of contested knowledge.

In the process, we have learned that as human beings, individually and sometimes in group settings, we are inconsistent. Individuals (and by extension institutions and groups) have been found to operate with serious cognitive limitations. A range of psychologists and behaviouralists posit that we use dual processes to make choices; we now characterize these as System 1 thinking, with fast, intuitive and emotional judgements, and System 2 processes, in which we engage in slow, rational and rule-based decision making. Behaviouralists Daniel Kahneman and Amos Tversky demonstrated that when we access System 1 thinking, we can be primed to seek or avoid risk in ways that are both changeable and inconsistent with standard expectations of how one should handle choice. Then, in our different roles as consumers, citizens and social beings, we apply different cognitive shortcuts that lead to biased and inconsistent choices.

As consumers, we are motivated, self-interested individuals seeking to optimize utility, in effect seeking choices in which the satisfaction exceeds foregone opportunities. What we choose depends on personal preferences, individual knowledge and capacity. In most settings, consumers have incomplete information, forcing us to make leaps of logic or to seek signals and advice from others. The information-deficit model asserts that more education and marketing will yield better outcomes but amplify or mute our concerns based on our experience and exposure to new events. In doing so, we fall prey to a number of cognitive traps, including: risk framing, where we tolerate more risk to mitigate losses and are risk averse when offered gains; affect, where we tend to be positively biased to things that trigger positive emotions (think fuzzy kittens); and mental accounting, where we compartmentalize our choices rather than normalizing risk and return across our life. As a result, we often present a muddled set of preferences to the market, willingly tolerating high-risk activities such as extreme sports, tobacco consumption and speeding on highways, at the same time as we push back against objectively safer but more uncertain choices, such as commercial air travel, nuclear energy or consuming GM foods.

We are similarly challenged as citizens in our dealings with the state. Conceptually, the state is contracted by its citizens to generate public goods and protect against social and market failures. Our policy and regulatory systems are designed to be evidence-based and non-partisan, using a range of decision rubrics, including the internationally accepted risk analysis framework and cost–benefit analysis. The goal is to avoid type I errors (i.e. avoid approving unacceptably harmful things) while at the same time minimizing type II errors (i.e. minimize rejecting otherwise safe and efficacious things). In effect, we authorize our system to process using System 2 thinking. The challenge is that the living, thinking, feeling humans who populate our governing systems are susceptible to all the cognitive biases that we see in consumers, with the addition of groupthink, overconfidence and confirmation biases. Computer learning and augmented intelligence are just beginning to transform how products are structured and designed, fundamentally altering the role that humans play in the research and development of new products. Our comfort zone regarding future digital technology innovations is likely to be radically altered.

Finally, people are multifaceted social animals. We belong to many groups that shape and motivate our values and beliefs (e.g. families, churches and advocacy coalitions), interests (e.g. unions and industry associations) and norms (e.g. our professions). Membership has its privileges, as it generates a sense of belonging. Social entanglements that engage our hearts and souls can motivate our reasoning, however, in ways that discount objective evidence. Anyone who is a sports fan knows that we lose our objectivity when discussing our team's performance in a match, especially in a loss that we can blame on a bad call by an official. We similarly exhibit political and social tribalism in the rest of our lives.

## Our Role in the Discourse

Each of us has engaged in the evolving discourse about new life science technologies and their application and use. We have crafted some of the evidence that sorts out the pathways these new technologies have followed, we have undertaken studies of how choices are made in firms, in the regulatory system, at the farm level and in consumer markets, and we have constructed some of the estimates of how much value these new products have generated and who has gained or lost along the way. For the most part, we let the evidence speak for itself. In the interconnected world we live in, however, we have been drawn into knowledge translation exercises. We have written opinion editorials, given public lectures and interviews, consulted with industry and government, produced reports (sometimes called grey literature) for think-tanks and policy venues, promoted our ideas and work through blogs and Twitter, and sat on national and international advisory boards and expert panels.

At one level, we recognize that the natural process of peer evaluation and scientific accumulation of our knowledge would be most fitting to the scholarly community that employs us. We also know, however, that if we don't talk about our work and its meaning, either it will be lost in the noise of ungrounded debate or, if it is noted, it will be misinterpreted and misused. So we accept that the strict norms of the modernist scholar are no longer enough. Our granting agencies and university employers agree – they encourage, motivate and judge us on our efforts to engage and translate our work to others.

The challenge is that, as we engage in the discourse beyond the science, our motives, efforts and impacts are open to multiple interpretations. Stuart and Peter, in particular, have been the focus of extensive scrutiny in terms of their engagement with industry and government. Each of us has to declare interests when we seek funding, do our research and publish our findings. Our publications, for the most part, aspire to the standards for the experimental essay set by Boyle: they are written in a plain, ascetic, unadorned (sometimes called functional) style, designed to provide adequate information to allow replication and to lay out the specific, narrow findings and implications of the experiment (also including experimental failures or exceptions). Some external advocacy groups, however, are concerned more with what they don't see than what we make public. The United States Right to Know (USRTK) group, in particular, wants to see the process, steps, missteps and machinations in our research process. The USRTK group has adopted the 'big tobacco' strategy of the 1960s and 1970s; they attack those that produce the science that quantifies the benefits and costs of GM crops and bioscience innovations in general. To do so, they have petitioned for one artefact of our efforts – our emails. Stuart and Peter together have had eight freedom of information investigations covering emails for as long as 20 years, one internal investigation of academic conduct and a judicial review in an appeal court. Although there are dozens of online, radio and newspaper reports of the suspicions of the advocacy groups, none of the disclosures or processes have found anything at all. Nevertheless, we remain suspect in the eyes of many advocates of alternate ways of doing research and conducting farming.

The rest of this book explores the contested science in agriculture, seeking to convey our sense of knowledge and meaning of its use and impact. Along the way we will highlight where the main disputes continue and how one might interpret them.

## Further Reading

Jasanoff, S. (2004) *States of Knowledge: The Co-Production of Science and Social Order*. Routledge, London.

Kuhn, T. (1970) *The Structure of Scientific Revolutions*, 2nd edn. University of Chicago Press, Chicago.

Mokyr, J. (2002) *The Gifts of Athena: Historical Origins of the Knowledge Economy*. Princeton University Press, Princeton, New Jersey.

Shapin, S. and Schaffer, S. (1985) *Leviathan and the Air-Pump: Hobbes, Boyle and the Experimental Life*. Princeton University Press, Princeton, New Jersey.

Thaler, R.H. and Sunstein, C.R. (2008) *Nudge: Improving Decisions about Health, Wealth, and Happiness*. Yale University Press, New Haven, Connecticut.

# 2 Expertise and Opportunism: Who Should We Listen To?

*We are experts. We each have earned doctorates in our fields, hold peer-reviewed tenured appointments at our university, have held research chairs in our fields and have collectively raised, as applicants or co-applicants, more than C$250 million of peer-reviewed funding. Jointly, we have produced more than 400 peer-reviewed journal articles and book chapters, written or edited 40 scholarly books, and been cited more than 7000 times by our peers. Most, if not all, of our work is relevant to this book. We decided to write this book because our expertise suggested to us that we have something that we should say to those beyond our normal audience.*

*Should you, the reader, choose to listen to us? For some readers, our self-proclaiming to be experts is a red flag that suggests caution when listening to our message – if not a signal to close the book and drop it in the nearest rubbish bin at the earliest opportunity. For the rest, our delving into the complex question of the role of experts in modern society, or for that matter whether such a thing as expertise actually exists, we hope can shed some light on a subject that has become a contentious issue in the governance of modern societies. We will focus on the question of the role expertise should play in the regulation of modern technology in the agriculture and food system.*

## Experts and Expertise

For some in civil society, expertise has become a bad word. It suggests an arrogance that leads to the presumption that those who do not possess it should defer to those who claim that they do. This runs contrary to egalitarian principles and smacks of class and hierarchy. The contrary is that everyone should make up their own mind about questions of the day and not have to defer to those claiming expertise. It is not the difference in the degree of knowledge that grates but the expectation of deference. It is the right to disregard the knowledge proffered that resonates. In the past, and some would argue the present, societies have been organized in hierarchies in which the norm was deference – to feudal landowners, to ordained clergy, to successful capitalists, to members of the judiciary, to police officers, to politicians. As our understanding of science and social science has expanded over the last 250 years, we have come to appreciate the complex world that we live in. Acquiring the scale and scope

S.J. Smyth *et al.* (2019). *GM Agriculture and Food Security: Fears and Facts* (S.J. Smyth *et al.*)

of knowledge to make sense of our complex circumstances is not a quick or simple process – most of us have neither the predilection nor the time to acquire such knowledge. The stock of knowledge in society has long exceeded the capacity or willingness of most individuals to acquire the full range needed to understand the complexities of our modern world. Hence, knowledge held by individuals is fragmented rather than comprehensive. In-depth knowledge is narrow and acquired by few in society. This creates an environment that encourages deference and, eventually, an expectation of deference.

Individuals can exercise their right not to be deferential and ignore the conclusions of experts. If one chooses to ignore the advice of fashion experts, probably the worst that can happen is that one's wardrobe does not convey the image one wishes to impart. Even in this case, however, ignoring the expert advice may lead to a lost job opportunity or exclusion from an event with a dress code. Although the use of clothing as a signalling device in job interviews may be unfair, and formal dress codes used as mechanisms to exclude participants are an imprecise and blunt instrument, the use of both is widespread. Heeding fashion experts can remove these particular barriers.

Although choosing one's right not to defer to experts can have relatively benign effects in many circumstances, in others the direct consequences and negative externalities can be detrimental and very large. Choosing to ignore the advice of experts regarding the link between lung cancer and cigarettes can lead to the detriment of an individual's lifespan being considerably shortened. The externalities are also large. If the smoker is the family breadwinner, the surviving family members may suffer poverty and reduced opportunities (e.g. an inability to attend college). If the healthcare system is public, there is an increased burden on taxpayers. If it is private, insurance premiums will rise, impacting all those in the insurance pool. If the cancer patient remains in the workforce for a period of time after contracting the disease, productivity may suffer. Although the reduced contribution of each lung-cancer sufferer may be small, it is still a non-zero reduction. On a larger scale, ignoring medical advice regarding the spread of AIDS in some areas of Africa has had significant negative effects on public health and economic productivity.

It is these types of externalities that lead those charged with the establishment of public policy to seek out those with expertise. It would be unconscionable for public officials not to seek out those with recognized knowledge of the science of engineering when planning a bridge, or architects when designing a building. A bridge collapse or a structurally unsound building coming down because those with expertise were not consulted or were ignored would lead to the condemnation of public policy decision makers. Politicians and other makers of public policy are seldom experts and, hence, must rely on input from experts in their decision making when complex matters are involved. This process should not be controversial, yet every day politicians are bombarded with suggestions, requests or demands that they ignore those with expertise. This raises the question, which knowledge of experts should be heeded, and which should be ignored?

One of the problems in answering this question is that the stock of knowledge is not complete and is not static. Thus, the knowledge of experts is fallible. Experts can be wrong. Experts cannot provide certainty. Depending on the state of knowledge of any particular subject, the probability of an expert being wrong may be greater or smaller. It is, however, very difficult for those without in-depth knowledge to determine the state of knowledge regarding any particular question and hence the likelihood of any particular expert being wrong.

At any particular point in time, decision makers may seek out a scientific consensus. Scientific consensus represents the ruling paradigm at that moment but is never a real consensus. It is some state of knowledge that is broadly agreed by the community of experts. Ruling paradigms can be wrong, however – sometimes spectacularly so. In the time of Galileo, for example, the paradigm was that the earth was the centre of the universe. It was stoutly defended. Galileo's work, along with that of others, eventually overturned this ruling paradigm. That the earth was flat was also once a ruling paradigm. Of course, scientific knowledge has increased manyfold since the earth was the centre of the universe and the world was considered flat. Current scientific experts probably believe that such spectacular overturning of ruling paradigms is remote, but in the light of such failures of ruling paradigms, it is probably not surprising that some members of the public question conclusions drawn by the scientific community of experts.

The idea that there is a scientific consensus pertaining to ruling paradigms contrasts with the reality of the scientific method that underlies modern science. The most that can be said is that there is a near consensus around a ruling paradigm. As discussed in Chapter 1 (this volume), the scientific method is based on questioning the hypotheses that underlie a ruling paradigm. This form of questioning is the basis of scientific progress. Often the questioning or alternative hypotheses brought forth come to nothing, but in other cases they are valid and the paradigm shifts. One of the consequences of the lack of a true consensus is that someone looking for a reason not to defer to a scientific consensus can almost always find a scientific expert that denies the validity of the ruling paradigm. For example, while there may be a near consensus regarding greenhouse gases as a cause of climate change, there are scientific experts who are climate-change sceptics or outright deniers.

This process is especially important for how we respond to a range of new biosciences. Although there may be a near consensus regarding the safety of human consumption of GM foods or the low risk to the environment of growing GM crops, there are also members of the scientific community who believe, if nothing else, that there is insufficient scientific information to allow such products into the human food chain or the natural environment. The absence of a consensus allows those with agendas to point to those scientists whose views do not align with the near consensus, to bolster their divergent views. Those with agendas that are aligned with the near consensus can cite scientists who accept the ruling paradigm. It allows for picking and choosing among experts.

How does establishing a near consensus take place in the context of public policy making? One example is international scientific organizations that are charged with devising internationally harmonized standards for food safety, animal health, plant health and the environment. The aim of harmonizing standards in these areas is the facilitation of international trade. Known collectively as sanitary and phytosanitary standards, they can, if not harmonized, act as significant barriers to international trade. A range of international organizations were initially established because some countries were putting in place trade barriers, justified on nefarious sanitary and phytosanitary grounds, to provide their domestic producers with economic protection from foreign competition. The three leading scientific organizations are the Codex Alimentarius Commission (established 1961) for food safety and human health; the World Organization for Animal Health, formerly the Office International des Epizooties (established 1924) and the International Plant Protection Convention (established 1951).

Although the means of arriving at an institutional consensus on a harmonized international standard vary slightly, they each follow a similar pattern. The organizations employ their own scientists to draft standards in particular areas of concern (e.g. maximum residual levels for pesticides or protocols for dealing with outbreaks of 'mad cow' disease). Once the draft standards are developed, they are released for comment by scientists designated by the governments of the organization's member countries – approximately 180 countries. These scientists are able to suggest changes to the draft standards and to react to those suggestions. Through this interactive process, a broadly acceptable standard can be agreed, or not. If agreed, it represents a near consensus.

These three organizations have been given an official role in international trade through the World Trade Organization (WTO) Agreement on the Application of Sanitary and Phytosanitary Measures Agreement (SPS). In the SPS, these organizations are recognized as the official standards-setting bodies for the purposes of international trade. This means that if a country chooses to adopt one of the near consensus standards as the basis for its trade regulations then those regulations cannot be challenged by other WTO members as being unfair barriers to international trade.

Even if a country's designated scientists have been part of the process of developing a standard at the international organization, it does not mean that the government of a country must adopt the near consensus standard as its regulatory standard. The SPS allows a country to choose a higher standard as long as it can provide a scientific justification for the higher standard. Hence, the views of scientists that are not part of the near consensus are taken into account in the international institutional architecture for international trade. These international standards, however, are a clear attempt to devise a method of finding what scientific experts agree upon.

One of the criticisms levelled against the international standard-setting organizations by NGOs and others in civil society is that they do not accept input from scientific experts who are not officially designated – often those that would reject the near consensus. Such criticisms are often also levelled against national efforts to arrive at a near consensus for domestic policy purposes. Governments often bring together a group of experts to examine particular questions of science to inform the public policy making process. For example, the European Union (EU) has assembled groups of scientific experts to provide opinions on such controversial topics as the human health risks associated with consuming beef produced using growth hormones, or the risks associated with the cultivation of crops developed using genetically modified organisms (GMOs). In both cases, the groups of scientists convened found the levels of risk to be acceptable, but not zero. As there were scientists who did not agree with the near consensus arrived at by the scientists chosen by the EU Commission, the Commission was accused of bias in the selection of scientists to serve as experts.

The important thing to realize is that a scientific consensus is not likely to exist if it runs counter to the scientific method that underlies scientific progress. While near consensus is possible, it is not inevitable. When it is arrived at, either officially or unofficially, there will be scientific experts that do not accept the conclusions of the near consensus. Given that nay-sayers will exist, the question then arises as to whether one should listen to them as opposed to those associated with the near consensus. This leads to an exploration of the process of becoming an expert.

## So What Makes an Expert?

Being an expert suggests that one has accumulated a considerable degree of knowledge on a subject relative to a non-expert or layman. Being able to accumulate in-depth knowledge does not imply that one is more intelligent than someone who has not accumulated it – only that one is better informed on that specific topic. Although it is possible to accumulate knowledge outside the institutional arrangements of formal education, spending time in the formal educational system is the most likely route for knowledge accumulation in modern society. Fundamentally, knowledge accumulation is a time-intensive process. In the beginning, the process is most often directed by teachers and professors who have already accumulated knowledge – they streamline the process of accumulation by establishing what is considered essential. Tests and examinations provide benchmarks pertaining to the degree of accumulation. The streamlining process, however, means that time is focused and that broader lines of knowledge accumulation are largely closed off. This leads to expertise being narrow and to the creation of what are known as disciplinary silos. Although those acquiring expertise are willing to expend considerable time in knowledge accumulation, time is not costless, thus few are able to pursue broad-based accumulation of knowledge – most people must eventually spend some of their time making a living from their accumulated knowledge stock. They will continue to spend some time in accumulation, but selectively.

The usual mileposts in the process of knowledge accumulation are a 3- or 4-year bachelor's degree, a 1- or 2-year master's degree, a 3–6-year doctoral degree and then, in the case of many disciplines, one or two appointments as post-doctoral fellows lasting approximately 3 years each. This represents a substantial investment of time and money in accumulating the knowledge considered essential to laying claim to having expertise. Of course, depending upon the discipline, the doctoral or master's degree may be sufficient to be considered an expert – but over time the length of the process to be considered an expert has tended to increase.

The process of knowledge creation, post-accumulation and consolidation, for the most part, follows a well-worn path. Research is undertaken and then put into a form that can be considered for publication. In most cases, this can be either articles written for professional journals or books. Prior to publication, they are reviewed by existing experts and, based on the reviews, may be accepted, rejected or require revision and resubmission. Academic careers depend on the degree of success in having one's research accepted for publication. One is often only considered an expert after having success in both knowledge accumulation and knowledge creation. Expertise is, as a result, typically deep and narrow.

We three co-authors of this book have roughly followed the path laid out above. One of us followed the progression exactly, with movement through the bachelor's, master's and doctoral education process with only a 2-year break after a bachelor's degree. Upon completion of a PhD, an academic position was immediately taken up and the phase of knowledge creation entered into, along with teaching and guiding students in their knowledge accumulation. Another of us spent considerable time in the private sector after completing a bachelor's degree, only returning to formal education later in life, first attaining a doctoral degree and then spending some time as a post-doctoral fellow and research associate in an academic department before formally joining an academic department as an industrially funded chair. The third co-author progressed through the educational steps – bachelor's, master's, doctorate – before spending more than a decade in industry and government. He then joined the ranks of academia and knowledge creation. Each of us has a successful track record of producing high-impact articles and books published by academic presses.

Each of us can be considered an expert in our respective fields: international trade policy, technology management and technology policy. We are sought out for policy advice by governments, the private sector, the scientific establishment and international organizations. We have combined our accumulated knowledge and knowledge generation efforts pertaining to GMOs which has yielded several co-authored books and a large number of published articles in academic journals.

We would be the first to admit that despite our expertise, our joint work is not without controversy. How could it not be, given the contentious nature of public policy pertaining to GMOs? We consider ourselves part of the near consensus on GMOs. In the case of GMOs, there are many who claim expertise who do not consider themselves part of the near consensus – and who would be likely to take considerable offence at suggestions that they were. So, who should one listen to? All we can claim is that all three of us have worked hard at both the accumulation of knowledge and the creation of knowledge. We have jumped through all the conventional hoops and over the hurdles associated with being an expert. Unfortunately, the time and effort spent accumulating and creating knowledge does not mean that we are always right. The process of accumulating knowledge does not lead to correct outcomes in a deterministic way. Just as those that accepted the paradigm of the earth being the centre of the universe in the time leading up to Galileo's discoveries got it wrong, there is always the possibility that, despite our best efforts, we have it wrong. Of course, those who have come to alternative conclusions could just as easily be wrong. One can never be sure that those with accumulations of knowledge have used that knowledge to correctly draw conclusions.

While it is not possible to know if an expert will be right in their interpretations derived from their accumulated knowledge, it is likely to be possible to exclude some who claim expertise. There are various ways that individuals attempt to signal their expertise and cues that those seeking expertise can use to select them. Evidence of time spent successfully accumulating knowledge is signalled by degrees obtained. Not all institutions where time can be spent accumulating knowledge are equal. Universities, for example, are not equal and considerable effort goes into providing ratings and rankings of universities and their programmes. In short, where one undertook the accumulation of knowledge matters. In a similar vein, where one holds an academic position and the nature of the position attained also matters.

Although degrees are often taken to be signals regarding expertise, there may be opportunities for obtaining the signal without accumulating (much) knowledge. Degree mills turn out individuals, for example, with PhDs who have not invested the time required to accumulate knowledge. Selecting an expert using a degree as a signal, without also knowing the quality of the university where it was earned, can be folly. This is equally true whether one is selecting an expert who agrees with the near consensus or one that is antipathetic to it.

Another important step in evaluating an expert is to determine whether the knowledge that has been acquired would inform the question at hand. It makes a great deal of difference whether John Smith, PhD, commenting on the technical safety of a nuclear power facility has a PhD in nuclear engineering or rural sociology. Remember, accumulated knowledge tends to be narrow so a PhD in nutrition is unlikely to have any specialized knowledge of pesticide management. It is fairly common to see, however, individuals espousing antipathy to the near consensus in a particular subject area, having accumulated knowledge in a different discipline. Accepting conclusions from such self-proclaimed experts is no better than trusting one's own judgement. One should always know where an individual's accumulated knowledge lies and not overly value expertise claimed in areas outside of it.

Exploiting the confusion over signalling devices, such as university degrees and knowledge accumulation, can be an effective tactic in the heated battles over hearts and minds that often surround ideas. Indicating that a person advocating a certain idea has, for example, a PhD—as a proxy for accumulated knowledge–when their PhD is actually in a different area of knowledge accumulation, can be used to bamboozle those that are the target of whomever is arguing in favour of the position or idea. A holder of a PhD in genetics supporting local food production on the basis of their having a PhD is simply cynical manipulation in the vein of the snake oil pitchmen of the past.

The practice is sufficiently common that there is an entire industry established for the purpose of providing soft academic credentials. The worst of these institutions are outright frauds that require no knowledge accumulation, simply the payment of a fee. While these phony degrees may not carry any weight with potential employers who make the effort to check – although some do not bother – they can have value in the case of advocacy. Tabloids and various websites are filled with advertisements that do just that: Anna Jones, PhD, says eating carrots improves the eyesight of pilots. Readers are unlikely to check, or even be able to check, on the validity of the PhD, as neither the subject area of the PhD nor the institution from which it was attained are identified. If pressed, an often elaborate and official looking degree certificate will be produced. In some countries, certification systems for degree programmes and educational institutions attempt to separate legitimate from illegitimate degrees but in the borderless world it is always possible to acquire a degree certificate from such suspect institutions as Cambridge City University or James Harvard University, implying connection with the top global institutions in the world despite being no more than a numbered post-office box somewhere.

Perhaps more common than outright frauds are soft degrees that require some effort to be expended in knowledge accumulation, but without the full degree of rigour required in more mainstream degree programmes. Such degrees often appeal to those who wish the cachet of a degree but do not have the time or inclination to pursue more conventional formal education. Some educational institutions make a point of targeting such potential students and tailor their offerings to employed individuals with convenient and more flexible delivery. After all, providing such educational options can be very lucrative. These degrees are often offered in what might be described as unconventional disciplines, which makes it harder for potential critics to question their legitimacy. There is, for example, broad agreement about what a PhD in economics or physics should entail, but who is to say what a PhD in food security involves? There are no hard and fast rules about which holders of such degrees should be considered experts. The absence of a deterministic relationship between degrees and expertise creates opportunities for exploitation; everyone should be wary of self-proclaimed experts.

## Prejudice Trumps Expertise

While it is easy to blame any of our misguided preferences on the advice of others (whether triggered by frauds or experts), we cannot let ourselves off the hook. Pogo, a cartoon character created by Walt Kelly, wisely concluded 'we have met the enemy and he is us', in parody of a famous quote by American naval officer Oliver Hazard Perry in 1813 after defeating and capturing British Royal Navy ships in the War of 1812. We are hard wired to think in ways that make evidence-informed decision making difficult.

There are a number of cognitive practices we all follow that we should be aware of and consider as we search for enlightenment. One is what is known as confirmation bias. Confirmation bias involves each of us choosing to listen to and accept information that supports our pre-existing values and beliefs. Devotees of the right-wing Fox News might be one example; many of us who don't watch Fox News' excited form of reportage are perpetually amazed that their audience believe some of the more extreme and unsubstantiated takes on news. Some of us put Fox in the same category of the much-ridiculed tabloid newspapers such as the *National Inquirer*, which one can find at many grocery check-outs – their signature would be screaming headlines about pictures of the Loch Ness Monster, UFOs, women having alien babies or the latest breakup of Hollywood celebrities. While most of us would smugly say 'that isn't us', we are not immune. We don't need to go to the extreme, to find confirmatory bias. We all make choices that affect our opinions in significant ways. Take for instance the readership of *The Guardian*, a left-leaning UK newspaper, and *The Telegraph*, a London-based, staunchly conservative broadsheet. Few Brits read both newspapers – they choose the paper that fits with their world view. In the case of a devotee of *The Guardian*, reading *The Telegraph* would simply be dismissed because the paper would not confirm one's prejudices. In the current era, with instant access to a vast range of information available through the power of the Internet and various social media platforms, confirmation bias is particularly important. While in theory the wealth of information provides an opportunity for people to gain a much more diverse and balanced perspective on the world, most of us explicitly choose, or are directed by unseen algorithms, to sample more narrowly. We by choice or by chance end up living in echo chambers that reflect our pre-existing views.

Not so long ago, the information sources used by experts were simply not available to the general public. Non-electronic means of publication (e.g. books and academic journals) were largely confined to university library collections where restrictions on public access often applied. Further, academic disciplines develop specialized vocabulary, which those who accumulate knowledge in the area spend time learning but the public do not. The public did not have easy access to the information regarding complex issues. As a result, in most cases policy makers could consult experts without public engagement or outcry. It was a cosy and comfortable world of insiders.

Moreover, in the era of non-electronic publishing, what constituted expert knowledge was tightly controlled as academic work and expertise, which is subject to the peer review process. Although the exact review processes vary to some extent by discipline, any book or academic article submitted to an academic publisher is passed on to recognized experts for review. If the reviews are positive, then the work can proceed to publication. If the reviews are negative, however, the work will not be published.

In economics, for example, acceptance rates are often 25% or less and some publishers in other fields accept fewer than 10% of the articles submitted. Academic publishers value their reputations and have an incentive to ensure the quality of their output. Reviewing is most often voluntary and unpaid. It is a process that most academics understand as an essential role in the knowledge creation process. This process has remained even as the publishing process has become increasingly electronically based. Physical and practical access to the output of the academic enterprise has become more open through electronic media – although effective access to the content remains problematic. Reading technical papers usually requires a certain degree of proficiency with the subject matter, including theories, analytical methods, empirics and, almost always, the highly specialized and stylized language.

The arrival of the Internet and associated information technologies has, however, removed many of the constraints on both accessing and producing information that the print-based system provided. The result has been an almost unfathomable increase in supply. Search engines make almost anything produced electronically accessible – much of it free. Access is gained by a few touches of the finger – or through a voiced request. Content that requires a fee to be paid can be ignored if one does not want to pay the fee – there always seem to be free alternatives. This is because, along with the removal of the constraint on access, the constraint on publishing information has also been removed. If one is willing to take the time to write something, it will find a home on the Internet. Search engines match supply of content with demand. In the print era, access to information was restricted but the quality was affirmed. That has largely disappeared. Now anyone can post anything.

Academic publishers now face significant new competition – at least from the perspective of the non-expert public. The advent of open-access publishing emerged with the Internet. Many funding agencies imposed open-access requirements for any publicly funded research; usually this involves the author paying the publisher a specific fee to publish the paper openly on the Internet – most often out of the grant that funded the research. These papers are still rigorously reviewed. A few well-established, credible journals and a number of new, quality journals moved into this space to provide this service (e.g. PLOS and PubMed). However, open-access publishing with its lower cost of entry (compared against that of print-based journals, which had to pay for printing and shipping), and the greater reach afforded by the Internet, attracted a completely new type of actor. There was an emergence of predatory journals (whose sole criteria for publication is the payment of a fee, with no further evaluation of quality), online vanity book publishers and a host of knowledge translators operating blogs and websites. Little, if any, of the work published on these platforms meets the norms of scholarship – review is illusory and the only real barrier to use is the fees charged by the publishers. As a result, our system is now swamped by vast amounts of material that has not gone through rigorous review. We shouldn't be surprised as the conditions for this were obvious: there has been a rapid increase in the number of academics worldwide, and each needs to publish to progress in their careers; the number of traditional journals was relatively static; and publishing in legitimate academic forums is, if nothing else, difficult. With the advent of these new platforms, there is a temptation and an incentive for scholars to forgo the more rigorous, more established academic venues for quick and easy publication. This may be risky for individuals, as academics and university administrations, when aware, will dismiss such publications and may penalise those who publish in them. Nevertheless, the explosion of new journals has led to a rapid expansion in the work claiming to be part of the accumulated knowledge. No one knows the full extent of the effort, but some report that in 2014 around 8000 predatory journals published an estimated 420,000 articles, equal to about 17% of the total 2.5 million articles published annually via all platforms. The average author paid about US$180 per article, which would generate approximately US$76 million in revenue for the predatory journal publishers.

Predatory journals understand the importance of legitimacy and work hard at acquiring some degree of implied legitimacy. The more effort that is put into discrediting predatory journals, the harder the journal publishers work to attain a degree of legitimacy. Jeffrey Beall, a librarian and researcher at the University of Colorado Denver, published one of the first lists of predatory publishers in 2010; after a series of threats and lawsuits, he took down his list in 2017. Others, mostly anonymously, have taken up the task of shining light on this part of the academic space. This is just a new chapter in the long struggle over rewarding quality in higher education.

What is most important for us is the increased access by the broader public. If predatory journals and online vanity presses simply polluted the academic literature, most scholars and experts could sift the approximately 17% that are bad, from the good work. When these articles pop up in your search, however, how are you, as a member of the general public, to know whether they have been published by a legitimate or predatory publisher? As suggested above, predatory publishers work hard at creating the illusion of legitimacy – sometimes using simple tricks. A legitimate publisher may have a title – *Journal of Economic Theory* – while a predator might creatively call theirs the *Journal of Economic Theories*, which to the undiscerning seems plausible. Even the three of us have been fooled at times by simply not reading a journal title correctly.

Beyond books and academic journals online, in recent years a myriad of institutions and individuals have made material available online through websites, blogs, social media and other archives. Although some producers of this material really don't care whether anyone believes them or not, as they may get paid by advertisers for eyeballs on their page, many of them hope that the readers believe they have some expertise and are worth listening to. Some have garnered enormous followings. Experts now vie with charlatans, the opinionated and celebrities for the ear of the inquisitive. Actor Dwayne 'The Rock' Johnson, actress Gwyneth Paltrow, Oprah Winfrey and model Jenny McCarthy have more influence on what some individuals will eat or what health regimes they will follow than Nobel laureates and global experts. Historian Stephen Shapin has argued that we have in some ways entered a topsy-turvy world, where those claiming the right to advise proactively disavow any expertise. It is a free-for-all which appeals because it fosters creativity and diversity of opinions, but it provides no guidance as to which of the many voices one should heed if you think, as we do, that evidence has any value in our decisions.

The way that individuals acquire knowledge has also changed with the arrival of social media. Prior to social media platforms, individuals seeking information had to be somewhat proactive. They had to search. In earlier times, they had to leave their homes and go to libraries. The advent of the Internet meant that they didn't have to leave their home; but they still had to think up search terms, enter them into a search engine and comb through the results of the search to get the information they were seeking. With social media, information comes unsolicited, provided by one's so-called social media friends: 'I know you are interested in climate change and I came across this article that I thought might interest you.' It comes from a friend, sometimes even a person you may trust. You may look at the information provided because you trust the source. The challenge is that you have been entirely passive in the process – you were not actively seeking the information. Your social media friend could be an expert, but most likely is not. You will have no way of knowing whether the information you have been sent was provided by an expert. You have looked at it because you trust your social media friend.

Social media networks are a mechanism for the purposeful spreading of ideas. All it takes is a willingness to engage. Anyone with a message they want spread, whether it be corporations, governments or NGOs, can and do devise social media strategies to influence opinions and behaviours. If done well, individuals don't even know they are being manipulated. The voices of experts are there, but largely indistinct from the cacophony of misinformation.

## Expertise and Science in Food Policy

Science policy and the acceptability of new technologies is one area where it is difficult to identify experts, which has led to considerable tension, fraying tempers and bad feeling. In some places, there is a clear battle for the hearts and minds of citizens and consumers. The intensity of these debates has led to considerable acrimony. The debate is to some extent no longer about the appropriateness of expertise and evidence and is instead about the total dismissal of expert advice. There is an avowed anti-science mentality among some advocates. In the battle for hearts and minds, discrediting science and expertise in general has become a popular and, we think unfortunately, successful tactic.

One area where the anti-science tactic has been manifest is over the use of biotechnology for food production and food consumption. Agricultural biotechnology, but not pharmaceutical biotechnology, has proved to be very controversial. It is a relatively unique issue because it has become a concern among four groups that tend to have strong preferences. First, individuals who were already concerned with the quality of the food they were eating (e.g. organics, hormone residues in meat, pesticides residues on vegetables, antibiotic residues in chicken) are engaged. Second, some individuals are motivated to participate because of their concerns about our environment. Third, individuals uneasy with the use of transgenic technology (e.g. those who feel that gene transfers between species are interfering with God's work as they could not take place in nature) have doubts. Fourth, some individuals are concerned about undue influence of large agribusiness firms in the food supply chain, given that most of the firms developing and commercializing GM crops have been large multinationals. The bringing together of these four groups over one issue has meant that it has become a lynchpin in the anti-science campaign. The technology has become a lightning rod for a considerable degree of angst regarding the agri-food system.

The anti-GMO movement was lucky in that it was able to foster angst among the broader society due to the use of simple but effective imagery. NGOs, aided by a willing press, early on began playing on the transgenic nature of the technology by characterizing its products as 'Frankenfoods', hence associating it with one of the most powerful monster images in literature and film. The term attracted attention to the issue – and the press, one suspects, found that it sold newspapers. Following on these monster images, anti-GMO protesters could dress as monster corn or mutant fish at public events and garner considerable media attention by doing so. We each have been at events, and even had our picture taken, with some of these costumed characters. Awareness that there might be issues with the technology became widespread among civil society, but more so in some countries than others.

In the debate over agricultural biotechnology, the near consensus of experts supported approving and using the technology, whether that was in the US, where attitudes among the public were more open to technology, or in the EU, where anti-GMO sentiment held greater sway. In spite of what you might have heard, the near consensus among the experts assembled to report on the technology in the EU, and most countries around the world, was very similar to that of experts in the USA. In the EU and a few other places, however, public angst regarding agricultural biotechnology was such that decision makers chose to largely ignore the conclusions of their experts and establish regulations to prevent adoption of the technology for either production or consumption. To bolster this stance, imports were also effectively banned.

As with any new scientific discovery, the anti-GMO campaigners could find scientists with accumulated knowledge in the field that disagreed with the near consensus, but they were few. There were also a host of experts whose accumulated knowledge lay in other areas that could be enlisted into the anti-GMO cause. Those with the trappings of the expert – holders of fake degrees – were also there, as they are in many contentious areas of public policy.

As a result, there were sufficient anti-GMO voices available, initially on the Internet and latterly on social media, that confirmation bias could be engaged in by those seeking information. In effect, the information provided by those espousing the near consensus could easily be ignored by those who had made up their minds.

One complication for the technology was that those experts who tend to agree with the near consensus are poor communicators. They tend to couch their message in technical jargon and scientific language that does not resonate with the public. Anti-GMO proponents tend to communicate through stories and narratives to which the public can better relate. Experts whose views align with the near consensus are generally not comfortable engaging in battles for hearts and minds and are ill-prepared to participate in those activities. They are trained to communicate their information to other experts. While they are sometimes called upon to provide input to legislators or other public policy makers in formal settings, even these venues can be difficult in terms of communication. Experts with views contrary to the anti-GMO position, hence, are often easy targets for their media-savvy and committed opponents.

Although the anti-GMO campaigners have a range of tactical advantages over experts espousing the near consensus, experts still exist and have the cachet that comes with standard measures of accumulated knowledge and institutionally granted prestige in the form of degrees and academic/research positions. Even with confirmation bias, anti-GMO campaigners cannot ignore their existence and their ability to wield influence. They at a minimum deny anti-GMO campaigners the ability to claim a consensus in support of their contrary views on the technology.

This is where things get messy. NGOs sometimes cross the line between playing the ball and playing the person. One effective approach to winning an argument is to damage the reputations of those you oppose, in this case those who are part of the near consensus. One of the ways that reputations can be damaged is to question the independence of experts. The funding of science is complex: research is conducted in a combination of public institutions such as universities, government scientific establishments and private corporations, using funds provided by the state, by charitable foundations and by industry. In the case of biotechnology, more research is undertaken in the private sector than was the case for past scientific breakthroughs that applied to agriculture. For example, the Green Revolution crops of the 1960s and 1970s were largely developed in public sector institutions – universities, government research stations and international research institutes, such as the International Rice Research Institute in the Philippines and the International Maize and Wheat Improvement Centre in Mexico. The latter are part of the Consultative Group for International Agricultural Research (CGIAR) that is funded both through charitable foundations and by individual national governments.

Early research in the area of plant genetics was largely publicly funded for two reasons. First, most crops grown by farmers could be used subsequently for seed: once a new variety was released and grown by farmers, part of the crop could be saved for next year's seed. As a result, it was very difficult for those investing in developing new crops to garner a sufficient return on their investment based on one year's sale of a new variety. If progress was to be made in using genetic potential it was going to have to come from the public sector. Hybrid maize was the exception that made the rule; given the poor performance of saved hybrid seed, most farmers were happy to buy new seeds each year.

Second, in most countries it was not possible to patent living material. The legal framework for intellectual property did not extend to newly developed varieties of crops. Thus, once released there was nothing to prevent other firms from copying a genetic-based crop improvement. Thus, private investors had no incentive to engage in research and development activities.

In the 1980s, with the potential of modern biotechnology on the horizon and, with it the spectre of large public-sector investments having to be made, governments were unwilling to shoulder the financial burden. As a result, patent laws were changed, following court cases, to extend the remit of intellectual property to living organisms in certain circumstances. Technological advances, including genetically moderated hybridization, also reduced the potential for saved seeds to continue to deliver the degree of productivity enhancement as generations progressed. As a result, the potential for selling new seed each year improved. The private sector enthusiastically began investing in agricultural biotechnology.

Many private-sector companies moved quickly to hire or contract many of the public-sector experts in genetics to do work in this area. In this situation, it was easy to question their independence. In the battle for hearts and minds it was easy to sow doubt about the willingness of private-sector scientists to be fully honest and transparent: were they covering up results that would negatively impact their employer's bottom line and which might threaten human health or the environment? Casting suspicion is easy, even (or especially) if there is no proof. It is impossible to prove a negative! The NGOs actively campaigning against innovations used this to their social media advantage, playing to the public's emotions about not being capable of proving things to be 100% safe.

Governments have also encouraged universities to seek funding from the private sector to extend their resources – but this simply opens the academy to questions of the independence of experts working in universities. Experts assert that there are adequate controls to ensure their ethical conduct: the existence of checks and balances in the scientific process itself, such as replicability and risks to reputation, should be enough. Experts tend to dismiss the accusations that scientists and social scientists are corrupt. Confirmation bias, however, allows one to choose the interpretation that one seeks – those antagonistic to GMOs tend to believe that experts are corrupt, often with no evidence or examples. As with any problem of this type, one cannot prove innocence.

Due to confirmation bias, however, one can find the same individuals being inconsistent in their approach to those espousing a near consensus. For example, individuals who condemn the experts that agree with the near consensus regarding GMOs, willingly accept the views of those supporting the near consensus on other topics, such as climate change. This is in spite of some well-publicized accusations that some supporters of the near consensus on climate change have done exactly what anti-GMO activists fear is happening in the biosciences: there have been a few cases where investigators have manipulated or edited data to support their warnings regarding climate change, one might assume because they have a vested interest in keeping a focus on climate change as it increases their research funding. The irony is that the individuals caught manipulating their climate science work argue that we should trust them because they are subject to effective checks and balances in the research systems – in effect the same defence used by experts in the biosciences.

## Are We Destined to Live Without Scientific Advice?

Once one believes that experts are corrupt, it is easy to adopt an anti-science position in the development of public policy. Removing experts' advice and their accumulated knowledge as an input into public policy leaves the policy formulation process open to lobbying based on belief or vested interests. Given the complexity of the issues that face public policy makers over GMOs, and a host of other technologies, this seems folly at best and potentially catastrophic at worst.

Society expends resources to allow the accumulation of knowledge by small numbers of individuals so that they can become experts. We can and should use their advice. We need their advice. Only the most naïve believe there are simple solutions to all issues. Dealing with complexity requires accumulated knowledge. That does not mean that one should choose to heed an expert's advice due to deference. One should choose to listen because there is unlikely to be a better alternative for complex issues. It is important to listen but not to slavishly accept. Winston Churchill affirmed from his experience as the wartime prime minister that it was valuable to have scientists 'on tap' but not on top.

Not to at least listen to the experts among us is abject folly, and a denial of our history. Accumulated knowledge is the basis for our modern existence and the quality of our lives. The answer to the question, who should one listen to? is complex, and hence answering it is difficult. Our goal here was to shed some light on the challenges – after all, that is the best that experts like us can hope for.

## Further Reading

Gibbons, M., Limoges, C., Nowotny, H., Schwartzman, S. and Scott, P. (1994) *The New Production of Knowledge: The Dynamics of Science and Research in Contemporary Societies*. Sage, London.

Jensen, M.B., Johnson, B., Lorenz, E. and Lundvall, B.A. (2007) Forms of knowledge and modes of innovation. *Research Policy* 36(5), 680–693. DOI: 10.1016/j.respol.2007.01.006.

Oreskes, N. and Conway, E.M. (2010) *Merchants of Doubt: How a Handful of Scientists Obscured the Truth on Issues from Tobacco Smoke to Global Warming*. Bloomsbury Press, New York.

Phillips, P. (2007) Governing knowledge. In: Phillips, P. (ed.) *Governing Transformative Technological Innovation: Who's in Charge?* Edward Elgar, Cheltenham, UK, pp. 143–176.

Shapin, S. (2007) Science and the modern world. In: Hackett, E., Amsterdamska, O., Lynch, M. and Wajcman, J. (eds) *The Handbook of Science and Technology Studies*, 3rd edn. MIT Press, Cambridge, Massachusetts, pp. 433–448.

# 3 The Technology Promise: How Can We Feed a World of More Than 9 Billion?

*The line-ups to get into the community hall for the fall suppers put on by various service organizations – religious and others – frequently stretched far beyond the doors. The delicious smells seeped out, easily reaching those standing in line.*

*Support for these common events in small towns across Canada is still substantial; what is particularly amazing is the seemingly endless supply of food that is served over the course of an evening. Typical menus where I grew up were reflective of Canadian Thanksgiving suppers that included turkey, dressing, potatoes, vegetables and gravy. While this was the core of the menu, it was always supplemented with additional items, such as cabbage rolls, sausage, buns and salads, all of which were homemade, using ingredients grown and donated by community families. I think that what amazed me most as a young boy was the incredible display of pies and desserts, making the choice of just one extremely difficult.*

*Events of this nature were successful fundraisers for the small community service groups. People volunteered their time to organize the event, many others donated food items to be prepared for the supper, while still others donated time to prepare and cook the vast quantities of food required to feed those who attended. Although there would always be a few items that would have to be purchased for an event of this nature, as not all the necessary turkeys would be raised on local farms, the majority of the items were locally produced. A lot of time and effort was donated, as buns, pies and desserts would have been baked weeks in advance and frozen until needed. People would spend hours peeling vegetables, making salads and organizing the collection and delivery of specific food items, as most community halls weren't large enough to hold all of the food required at one time.*

*While sourcing and securing enough food to feed several hundred people in a small community requires a great deal of hard work by local organizers, this is dwarfed by the effort required to pull together to improve food security at a global level. Research, development and production efforts require international coordination, substantial funding commitments and functioning regulatory systems capable of approving and governing the food that is developed and consumed.*



S.J. Smyth *et al.* (2019). *GM Agriculture and Food Security: Fears and Facts* (S.J. Smyth *et al.*)

## The Current State of Global Food Security

Food security has recently gained renewed global focus due to the challenges of feeding a projected world population of more than 9 billion people by 2050. Rising incomes combined with population increases are forecast to double global demand for food, with some estimates suggesting that as much food will need to be produced in the coming 50 years as has been produced by mankind over the past 10,000 years. Meeting this increase presents a global challenge for the agri-food system. Currently the global agri-food system involves more than 1.3 billion farmers and millions of firms, with some element of the effort in every country. The remarkable achievement of the past century, however, is the organization of a global system of research, input industries, production, processing, transportation, storage and distribution. In this past century, we have successfully reduced food insecurity, from the risk of hunger affecting more than 60% of the world population in 1900, to affecting about 12% now. Admittedly, a larger portion of the population has episodic micronutrient deficiency, which is not new but suggests that we still have a long way to go to achieve consistent access to adequate food for all.

Nevertheless, there are hundreds of millions of people that currently do not enjoy an acceptable level of food security. A report by the United Nations Food and Agriculture Organization (FAO) in 2006 estimated that about 850 million people were food insecure in 2006. The report forecast that the food-insecure population would decline by one-third to just under 600 million by 2015, but the stark reality is that food insecurity is virtually unchanged after a decade of effort. In 2017, an estimated 820 million people did not have adequate access to food, and a further 2.1 billion people were classified as micronutrient deficient. The FAO suggests that, in order to meet growing food demand, agricultural production needs to increase at least by 70% worldwide by 2050, and by almost 100% in developing countries. Complicating this, climate change is projected to make production in some regions more variable and uncertain.

National self-sufficiency is not the answer. Rather, food trade does and probably should continue to contribute to food security, as trade is pursued out of both absolute need and relative benefit. A country such as Canada, for example, could grow many fresh fruits and vegetables but imports most of these products from abroad, partly because the product is cheaper and partly because of the seasonality of fruit production in Canada due to its northern climate. This does not mean that Canada is dependent on imports. Other countries, such as Egypt, on the other hand, are dependent on wheat produced elsewhere to feed their growing population – they simply don't have sufficient fertile land and other resources to meet their domestic needs. The degree of interdependence in effect comes down to whether a country could feed its population with its own production using its own resources, not whether it actually does so.

The FAO estimates that very few countries are truly self-sufficient. The only countries to unambiguously have the capacity to produce more food than they consume are Argentina, Australia, Burma, Canada, France, India, Russia, Thailand and the USA, along with a few other small countries. Approximately 16% of the global population currently relies on imports of food produced elsewhere for their basic food security.

There are three major elements to food security: (i) food availability; (ii) sufficient income to acquire food; and (iii) over the long run, increased agricultural productivity and supply. An individual is food insecure if safe food is not available on a daily basis. Even if food is available, individuals may remain food insecure if they do not have sufficient income to purchase it. As global populations expand, ensuring that enough food is available and affordable requires higher agricultural productivity. Otherwise there will be more individuals chasing ever scarcer food, leading to higher prices, lower availability and rising food vulnerability. The FAO has estimated that crop yields will need to increase 2% per year to feed the forecast global demand. Over the past 20 years, yield increases have averaged 1% annually for wheat and rice, two critical global staple crops that underpin food security. To be able to develop plant varieties that are capable of providing higher yields and pushing back the Malthusian prospect of population growth outpacing food production, no safe and effective plant breeding technologies should be summarily dismissed.

Scientific breeding has certainly done its part. For thousands of years, farmer selection isolated valuable crops and tailored specific strains of natural crop species – landraces – to specific ecosystems and uses. With the advent of modern plant breeding where close relatives of plants were forced to cross-fertilize, yields started to accelerate. The single biggest historical innovation was the invention of hybrid maize, which transformed maize production, expanding area of land planted to maize and more than doubling yields over the first 50 years of adoption. More recently, the cracking of the double helix, the definition of the role of DNA and RNA in heredity, refinement of mutagenic techniques, and the subsequent invention of marker-assisted breeding tools, genomics sequencing, gene isolation, transgenic gene-splicing tools and now gene editing, has pushed back the natural limits of plant reproduction. As a result, a whole host of new traits and applications are possible, including herbicide tolerance, insect and virus resistance, quality traits and a range of abiotic attributes, such as sunlight and drought tolerance. Scientists and entrepreneurs wax eloquent about the endless frontier of possibilities.

While science and technology have great potential to generate new, more productive crops and animals to meet growing and differentiated demand, the real challenge will be to ensure that the governance system is able to translate those opportunities into better outcomes. To a significant degree, this is a two-pronged problem. First, regulatory frameworks are complex and difficult to navigate for technology developers; recent developments have worked to increase complexity, time and cost to receive commercialization approvals, arguably without any corresponding change in safety. While commercial developers of large-scale crops in developed nations are usually able to navigate these complex processes, proponents of innovation for small-scale subsistence crops in developing countries are finding it difficult and expensive to engage in this process. Second, international commodity trade, which translated many of the earlier gains in productivity to food insecure people around the world, is in danger of returning to the world of 100 years ago when food trade was severely restricted. If global food security is truly going to be enhanced, coordinated research, regulation and trade actions are desperately required. One area where these challenges are manifest is in the international regulation and trade issues surrounding modern agricultural biotechnology (agbiotech).

There is considerable evidence of serious underinvestment in agricultural research and development (R&D) over most of recorded history. Even if investment could be increased to close the current shortfall, there are considerable lags – often more than 25 years – between making investments and full realization of productivity increases. A good portion of this lag is due to the 12–15 years that it takes for plant breeders to develop, field test and receive regulatory approval for new varieties. Many reasons exist for underinvestment in agricultural research, including the following: the inability of breeders to capture enough benefits to justify commercial investments; resistance to technological change; poor intellectual-property protection; high costs in identifying and acquiring existing intellectual property; inadequate public investment in plant and animal breeding; uneven application of technologies, which limits the size of the market; and long and costly regulatory processes for new technologies.

The risk assessment process is one of the variables that is within the scope of each state to sort out. Prior to commercial production, products must be reviewed to ensure that they do not pose an unacceptable level of risk to society. In the period after the Second World War, and up until this century, national and global regulatory groups worked to establish the risk analysis framework, with a set of principles, protocols, processes and procedures to assess, manage and communicate the scientific basis for risk evaluation. That system is formally embodied in most national regulatory systems for most products, including foods, and is the basis for international cooperation and management of global risk. Over the last 20 years, however, the process of risk assessment has been evolving and diverging geographically. The major spur for the diverging treatment of different foodstuffs has been agbiotech and the commercialization of GM crops. The main divergence is not in the science but in the social response. The USA, Canada and a range of GM producing and exporting countries have largely treated GM foodstuffs as similar to conventional foods, in that the focus has been on assessing any differential effects of GM traits in toxicity, allergenicity or environmental sustainability. The EU and a range of other food-importing countries have largely accepted the results of the scientific risk analysis but have asserted that their consumers and nations have the right to selectively approve foods based on the social acceptability of any embodied technologies or production methods. Although the WTO generally disapproves of measures that differentiate between goods that are essentially alike except for their production and processing methods, neither it nor any other international agreement can restrain or compel national policy; they can only moderate or manage resulting disputes. The conflict over agbiotech is often portrayed as a disagreement between the EU and the USA and, while they have been major champions of the divergent approaches to risk, the rift has global implications for investments in agricultural technologies and food security.

The increasing uncertainty and the divergent decisions of regulatory systems are reducing the appetite – both public and private – for the required investment to bring products that use new biosciences to the market. This has a number of effects. In the first instance, increasing and diverging regulatory hurdles are raising the costs and extending the time to achieve compliance, which raises the overall costs of bioscience innovation of foods. Second, the rising probability of failing to demonstrate a low enough level of risk to satisfy regulators generates uncertainty, which forces investors to discount returns in their planning models. Third, in most cases only a subset of potential markets accepts a new bioscience food, delimiting and truncating the market. All of this inhibits investment in R&D, reducing the rate of technological change just when it is most needed to ensure food security.

South Africa provides an excellent example of the importance of incorporating all available tools for improving food security. In South Africa, the average small landholder cultivates only a few hectares. Given the small size of the fields, weeding is done by hand. To prevent weed damage, conventional maize requires 276 h of hand weeding per hectare, sorghum 150 h/ha, cotton 200–400 h/ha and rice 200–418 h/ha. Research undertaken by Marnus Gouse at the University of Pretoria found that smallholder farmers in South Africa who had adopted GM maize spent on average 267, 177 and 152 h/ha on manual weeding over three seasons, down from the average 276 h/ha for conventional maize. Weeding is typically done by women and children. Female household members indicated that an average of 10–12 days of effort saved in their GM maize fields enabled them to spend more time at home, caring for children or in their vegetable plots while the children were able to continue their studies uninterrupted by weeding duties.

## Economic, Environmental and Human Health Benefits of GM Crops

The global food system is only part way towards a major transformation. Since the Second World War, the food industry, aided by and partnered with governments around the world, has expanded supply sufficiently to reduce the incidence of malnutrition and food insecurity, in spite of a rapid rise in the global population. Despite this progress in expanding supply and meeting the increased demand for food, the combination of slowing agri-food productivity growth, widespread soil depletion, increasing water shortages and the effects of climate change all threaten the achievements to date. As if this is not enough, demand is inexorably rising, both due to increasing global population and to higher incomes that translate into increased demand for animal protein, which accelerates demand for feed and competition for food crops.

The 2019 harvest will mark the 25th year of commercial GM crop production and while numerous environmental non-governmental organizations (eNGOs) continue to assert that there are no benefits from biotechnology or GM crops, the evidence suggests that GM crops contribute significantly to improved agricultural productivity and food security. A wide range of positive economic, environmental and health benefits have been realized from GM crop adoption, distributed among both adopting and non-adopting farmers, consumers and countries around the world. By some estimates, this wave of technological innovations has generated global net benefit of more than US$186 billion so far and continues to deliver large benefits annually.

Economic benefits are widely reported, with increased yields from lower abiotic and biotic stresses, as new varieties better control for weed and insect pressures. In the case of GM cotton, yield increases of 37–58% have occurred in India. In China, GM cotton raised adopter income by US$500 per hectare. In South Africa, GM cotton yield increases of 89–129% and raised net incomes equal to 2–4 months of wages have been reported. GM maize adopters in the Philippines have a net household income that is 50% higher than non-adopters due to the ability to secure off-farm employment following the adoption of GM maize. GM cotton adoption in India raised vulnerable household incomes (those defined as living on less than $2/day) by 134%. Finally, farmers in Brazil report that in some crop-producing regions, farmers are able to produce three crops per year due to the innovations provided by GM varieties.

The distribution of economic benefits varies widely between farmers, consumers and innovators. Studies that have attempted to measure the cumulative benefits of a GM crop range from the assessment of individual countries to global assessments. Cumulative economic benefits ranged from US$2 million from GM maize in Spain to US$160 million and US$315 million for GM cotton in the USA and India. GM canola in Canada has increased farmer incomes by US$266–304 million annually. GM maize in Argentina increased income by US$5 billion between 1996 and 2010. The global economic benefits for GM soybeans have now reached US$46 billion, with 31% estimated to accrue to consumers of soy and related products, whether they consume the new varieties or not.

We can unambiguously assert that with the adoption of GM crops over the past two decades, farmers must be deriving significant benefits from the technology or they would have begun to return to previous crop options – which they have not. In every country that has allowed unrestricted commercialization, adoption rates by crop have exceeded 75% of total acreage within 3–5 years of introduction. A study on the economic benefits of GM maize adoption in the USA, found that GM maize created US$6.8 billion in extra value annually, with 60% going to non-adopters due to lower insect pressures on adjoining non-GM crops. Graham Brookes and Peter Barfoot have examined the economic impacts of GM crops annually over the past decade. Their most recent report on the 2016 crop year estimated that the annual global economic benefit of GM crops exceeded US$18 billion. Cumulatively, over the period from 1996 to 2016, the authors estimated that the economic benefits of GM crops exceeded US$186 billion. The production increases from GM crops have made a substantial contribution to improving global food security, generating an additional 213 million tonnes of soybeans and 405 million tonnes of maize over the first 20 years of adoption – output that would not have been produced without the technology.

Even though farmers appear to gain, some are concerned that the new crops could hurt the environment or consumers. The available evidence of GM crop safety and efficacy, however, is compelling. An analysis of 147 studies detailing GM crop impacts found that while crop yields increased 22% and farmer profits increased 68%, chemical pesticide use dropped by 37%.

Environmental benefits are increasingly being observed through the reduction in herbicide, pesticide and insecticide applications to GM crops. Pesticide applications have dropped dramatically with GM cotton adoption. The first studies examining the environmental benefits of GM crops were undertaken following the adoption of GM cotton in China in 1999, finding that the adoption allowed farmers to spray less frequently, in some instances dropping from 30 applications per season to 3, but more commonly from 12 to 3 or 4 applications. Insecticide application in China dropped from 14kg/ha to 4kg/ha. In Canada, the environmental impact of GM canola resulted in a reduction of 53%, or 1.3 million kg of herbicide active ingredient annually. There has also been a corresponding increase in conservation tillage practices, especially in North and South America. Conservation tillage practices reduce the amount of carbon released from tillage, increase the moisture content of the soil and reduce soil erosion. In Canada, conservation tillage rose from 11% to 64% in canola production between 1999 and 2006, while in the USA its use increased from 30% to 63% with GM soybean production.

One way to estimate the environmental costs of not adopting GM crops is by investigating the ban on GM crops that various Australian states had in place after GM canola was approved by the national government in 2003. In 2004, a country-wide moratorium on planting GM canola was implemented. It was not until 2008 that the central canola producing states of New South Wales and Victoria lifted the moratorium, followed by Western Australia in 2010. An examination of the environmental costs of the Australian moratoriums showed that delayed adoption of GM canola production cumulatively resulted in the application of an additional 6.5 million kg of active chemical ingredient, about 14% more than could have been applied if herbicide tolerant varieties had been adopted at a rate similar to that of Canada. The application of these additional chemicals involved an additional 7 million field passes, requiring 8.7 million litres of diesel. Cumulatively, an estimated 24 million additional kg of greenhouse gases were released due to the delay in adoption of GM canola. The additional input costs, coupled with lower potential yields, forced Australian canola growers to forgo US$355 million in net income due to the moratorium.

While scientists and courts continue to debate whether GM traits or the chemicals that they are paired with are completely safe, the evidence is that they are unambiguously safer than the alternatives. Significant health benefits to smallholder farmers and farm labourers due to reductions in pesticide applications resulting from the adoption of GM cotton have been reported in India, China and South Africa. The adoption of GM cotton in India lowered the number of reported pesticide poisonings among agricultural workers by somewhere between 2.4 million to 9 million cases per year. In Burkina Faso, GM cotton resulted in 30,000 fewer cases of pesticide poisoning among farm workers annually. The recent abandonment of GM cotton in Burkina Faso due to short cotton fibre lengths and low industry demand will force cotton farmers to resort to the use of more toxic chemicals, with a concomitant rise in pesticide poisoning among farm workers.

Many eNGOs, including the self-acclaimed advocate of India's poor, Vandana Shiva, continue to advance stories about the unfortunate situation of suicides among farmers who have adopted GM cotton. However, Guillaume Gruere and Debdatta Sengupta studied the situation and concluded that there was a one-third reduction in the suicide rate following the release of GM cotton among Indian farmers, based on extrapolating the suicide rate before GM cotton and the lower observed rates after commercialization. The increased profitability of small landholder farmers growing cotton was found to alleviate some of the long-standing pressures on farmers, resulting in fewer, not more, suicides.

Meanwhile, significantly better human health outcomes for consumers have been linked to the reduction of insect-related toxins in GM maize. Studies have shown that insect damage to the ears of maize was reduced by about 60% through the use of GM maize and the resulting crops were found to have 29% fewer mycotoxins, 31% fewer fumonisins and 37% fewer thricotecens. The effects of these toxins range from being mildly annoying to humans, to them being the cause of some types of cancer, most notably metastatic oesophageal tumours.

The evidence continues to mount. GM crops have been demonstrated to increase production, to reduce inputs such as chemicals applied to control insects and weeds, to benefit the environment through the reduction of tillage as a form of weed control and to provide health benefits to farmers and consumers. In 2016, the American National Academy of Science undertook a detailed and rigorous assessment of over 900 scientific publications on GM crops, concluding GM crops to be safe for both the environment and humans.

With the abundance of evidence in support of the substantial benefits provided by GM crops, why is there such a strong backlash by the environmental movement?

## Opposition to Bioscience Strategies for Global Food Security

While the science capable of increasing food security continues to make advancements, the politics pertaining to technology options for feeding a future population is increasingly hostile. Billions of dollars are invested annually into the development of new crop varieties. Governments have concluded that it is too costly for taxpayers to fund all possible research solutions and, as a result, private corporations have stepped forward to fill the gap. The resulting public–private research partnerships have come at the request of governments and have successfully leveraged public R&D investments with additional private-sector funds. This research is designed to achieve numerous breeding goals, including improved drought tolerance, insect resistance, larger root structures, general yield gain, nutrient improvements and the potential to reduce crop spoilage. All forms of plant breeding are being used to tackle plant variety improvements, including conventional plant breeding and applications of modern biotechnology.

The challenge that plant breeding and agriculture in general face is that the opposition to innovative means of plant breeding simply rejects these options for improving food security. Major eNGOs argue that organic crop production will be able to feed the world instead. Simply put, this is naïve and misguided. While organic farming generates real value to a segment of consumers and some economic and environmental returns to organic producers, the evidence indicates that relying solely on organic agriculture to feed the world will only increase food insecurity, starvation and malnourishment, because organic crops yield less than either conventional or GM crops.

When we review relative yields for organic crops compared with the conventional alternatives, it appears that organic rice yields are nearly 40% lower. Wheat fares a little better, with organic spring wheat yields 34% lower, and organic winter wheat yields 29% lower, than existing conventional varieties (there are not yet any commercial GM rice or wheat varieties to compare against). Organic maize yield variations depend on the intended use, with organic grain maize yield 35% less than GM maize and organic sweet corn yields 16% less than the GM varieties. Relative organic yields are even lower for crops such as nuts, fruits and vegetables. The yield differences are as high as 60% for tomatoes, 70% for spinach, 60% for strawberries and 67% for cranberries. Organic horticulture crops, such as nuts and fruits, also require greater supplies of water, which will be an important issue for many producers in the future.

The FAO measures food security predominantly through what are known as the staple crops: maize, rice and wheat. In many countries, the major proportion of daily nutrient consumption is met through these crops and often only one of the crops contributes the majority of the diet. For example, in many Asian countries up to 80% of daily nutrient intake for some at-risk subpopulations comes from eating rice.

If the controversy about production choices remained with scientists, producers and consumers, the market could determine what gets produced, where and in what quantities. Governments, however, have engaged and confounded any market resolution. The EU, in particular, has frustrated efforts to seek a resolution. The EU was initially accommodating, if not enthusiastic, about GM crops, but by 1998 the system of union-wide review and approvals had stalled. The halt to further use was sustained by EU parliamentarians, even in the face of contrary opinions from the European Food Safety Authority (EFSA) and a 2010 European Commission review of 130

EU-funded biotechnology research projects, which concluded that GM crops were as safe as conventionally bred crops. After almost a decade of inaction and a judgement at the WTO against the de facto moratorium, the EU finally crafted new rules that in theory would allow review and approval of GM crops, provided they met all the health and safety standards, were produced and marketed in ways that provided consumer labelling of GM ingredients and sustained a system that would allow farmers to produce and market conventional crops to those seeking GM-free foods. The resultant labelling and coexistence regimes were judged by the biotechnology companies to be onerous but workable. The resultant system, however, had effectively decoupled the scientific risk assessment and product approval processes within the EU. The EFSA conducts the risk assessment using science-based methodologies and provides a report of its assessment, which can suggest approving the commercial production of a GM crop to the European Commission. The actual product approval process resides with the Standing Committee on Plants, Animals, Food and Feed of the European Commission, which has consistently failed to approve GM crop varieties for commercial production. Then, in December 2014, the European Commission announced an agreement allowing member states to ban GM production within their national territories. This resulted in 19 of the 28 EU member states immediately enacting GM crop bans. Moreover, in the face of shipments with detectable non-approved GM traits, the EU reaffirmed its zero-tolerance threshold for even trace amounts of GM traits that are fully approved in, for example, Australia, Argentina, Brazil, Canada or the USA. These GM traits usually will have had risk assessments cumulatively undertaken by over 200 federal regulators at that point. Somewhat ironically, the EU has higher tolerance for such unappetizing items as insect fragments, sticks, rocks, manure and animal parts. In all, eight EU member states have planted some GM crops at some point since 1995, but in 2018 only Spain and Portugal planted small amounts of GM maize.

Investment decisions of individual firms are determined, in part, by the size of the potential market for the products of the new technology. When countries ban new technologies or make approvals sufficiently risky, then the potential market shrinks. This reduces the likelihood of a positive investment decision. Meeting the goals of improved global food security will require a range of efforts, but a central element is improving agricultural productivity. Agbiotech can be a major contributor to advances in agricultural productivity. Any major increases in the riskiness of investments in R&D will reduce investment and the rate at which agricultural productivity increases. It is those countries needing to increase their agricultural productivity that suffer the most when investments in biotechnology are inhibited.

The effects on food insecure countries, particularly in sub-Saharan Africa, are more direct. The EU prohibits imports of all unapproved GM crop varieties. Shipments are refused when minute traces of GM material – either the mingling (or sometimes called comingling) of GM with non-GM varieties of the same crop or a different crop – are detected. Hence, if African countries exporting to the EU adopt productivity-enhancing GM crops they risk losing their EU markets. Given that absolute segregation of GM and non-GM crops cannot be assured in the commercial system (in developed, much less developing, countries), African countries eschew the adoption of GM crops, thus forgoing improvements in their agricultural productivity. The failure to adopt GM crops in the EU feeds back into global investment decisions – one result is that there is very little investment into the adaptation of GM crops suited to agronomic conditions in developing countries, especially tropical crops.

## The Future for Food Security

Food security is more uncertain now than in any recent decade. After more than 70 years of increasing trade liberalization and globalization of markets and significant success in reducing food insecurity, there seems to be a pause that threatens to both destabilize the global world order and knock back the effort to mobilize science to create a more sustainable and food-secure world. Governments, industry and citizens are pulling back from using competitive markets to promote their interests, raising fears that we are moving into what some have coined a 'neo-medieval' world of special interests.

Concerns over the effects on human and animal health from the consumption of GM food crops, as well as worry about an array of potential impacts on the wider environment, have divided public opinion over the merits of using the technology. One consequence is that governments have, for the most part, been very cautious about accepting GM crops for cultivation and consumption. Instead of relying on the global system to ensure safety and value, most nations have renationalized the regulatory process, so that developers now need to secure approvals for each intended use in each national market. An effect of the expansion and duplication of regulatory oversight is that the costs involved (and time spent) in achieving regulatory approval for the commercial release of a GM crop have risen. One result is that public-sector researchers and universities have been limited in the degree to which they can undertake research in this domain. A range of studies show that there are significant costs to gaining regulatory compliance – product attributes, market context and technical aspects of measurement all have significant impact on the estimated costs of regulatory approval.

While the opportunity to use GM technologies in the EU seems to be lost, there is a larger opportunity that the world cannot afford to waste. A host of new plant breeding techniques and technologies are poised to enter the crop variety development sector. They offer an exciting opportunity to reverse the trend to weaker yield growth. Although many of these new technologies and techniques do not involve cross-species manipulations – cross-species manipulation or transgenics was the major objection initially raised regarding GM technology – there is a movement afoot to outright reject these technologies, especially in many of the EU member states that have exercised the opportunity to renationalize decisions about GM crops. The opposition of eNGOs in France to new mutagenesis plant breeding technologies resulted in a Court of Justice of the EU (CJEU) ruling in July 2018 that genome-specific mutagenesis must be regulated as transgenic plant breeding. Targeted mutagenesis involves no foreign transfer of DNA between plant varieties, but the CJEU chose to accept the legal arguments that eNGOs opposing the technology had put forward rather than the scientific evidence coming from global regulators. As a result, a real opportunity to accelerate crop productivity, lower the ecological footprint of food production and improve the lot of many food-insecure farmers and families around the world may have been foregone. The burden of increasing global food security cannot rest in the hands of the few countries adopting GM crops, led by Argentina, Australia, Brazil, Canada and the USA, but must be shared equally among all countries. The commercialization of new, higher yielding crops needs all leading food producers to engage. The EU stance is unconscionable – it has a duty to use its wealth and resources on behalf of humanity and not only its own parochial interests.

The advent of biotechnology products has triggered a spirited debate about how we should assess risks, who should undertake the task of defining what is acceptable, what rules we should draw upon and where we should vest the authority to decide. Some eNGOs have even gone so far as to suggest that unless science can guarantee zero risk, then the innovative technology should be banned. In the recent past, there was a strong move to normalize and institutionalize a 'science-informed' system in national regulatory systems, international science and international treaties. Instead, there has been a substantive pushback against the 'privileged' role scientists and science institutions play in decision making, especially pertaining to the introduction and use of GM crops. The underlying information, valuation and rules for deciding what gets used or discarded are the focus of two discrete groups. The modernist, positivist community is advocating science-informed decision making and rules while some so-called post-modernists want to incorporate socio-economic considerations to deepen the contextual base for decisions. Either approach can lead to regulatory capture but the methods that would be involved in exploring socio-economic considerations are less well developed and codified than the scientific fields, which leaves them more open to capture and manipulation by vested interests. This would just compound the already difficult challenge of securing adequate nutrition for the burgeoning world population.

In the past few years, there is increasing evidence that the post-war consensus supporting a globalized, harmonized, multilateralized economic and political system may be unravelling. Whereas the calls in the 1970s were for a New International Economic Order to bring greater equality to the global economic system, our view is that we are facing pressures that could lead to a New International Economic (Dis)Order. The evidence is scattered, but one can see pressures for disintegration and isolation emerging among governments, in industry and among citizens around the world.

In the industrial setting, the biggest change that we already see is in the number of firms willing to invest in advanced technologies in the food industry. In the early days of the development of biotechnology in the food industry, all the large national and multinational seed, chemical, handling, production and retail enterprises made efforts to assess the potential value from the technology. Meanwhile, a diverse range of scientific and entrepreneurial startups contested the industry, handsomely financed by the global venture capital industry. As controversies erupted and markets segmented, many of the large firms pulled back and retrenched in the conventional food space, while the remaining actors consolidated, to the point now that there are only three main commercial agri-food biotechnology companies (DuPont/Pioneer/Dow, Bayer/Monsanto and BASF) and one major state-sponsored company (the merged ChemChina/Syngenta). Few if any start-ups are testing the market for GM foods. Instead, the entrepreneurial focus has shifted to data management and precision farming.

Normally we would take a deep breath and hope that the wisdom of crowds would see us through. However, it is far from clear whether collective intelligence is still functioning. The dominant role the Internet, Twitter and blogs played in the UK referendum on Brexit and the 2016 US presidential election has raised the prospect of a 'post-truth' future, where facts, evidence and thoughtful discourse are rejected or ignored. In such a world, special interests have much greater opportunity to spin messages that convert our world into what political economist Robert Gilpin has called a neo-medieval system where we are bound by a complex web of inward-looking, narrowly focused, vested interests, analogous to the guild system of the Middle Ages.

We can and must do better than that with all the resources and capacity at hand. The millennial goal of food security for all is within our grasp if we can organize ourselves effectively.

## Further Reading

Brookes, G. and Barfoot, P. (2016) Global income and production impacts of using GM crop technology 1996–2014. *GM Crops & Food* 7(1), 38–77. DOI: 10.1080/21645698.2016.1176817.

European Commission (2010) A decade of EU-funded GMO research 2001–2010. Directorate-General for Research and Innovation, Brussels. Available at: http://ec.europa.eu/research/biosociety/pdf/a_decade_of_eu-funded_gmo_research.pdf (accessed 4 February 2019).

Food and Agriculture Organization (2012) *The State of Food and Agriculture: Investing in Agriculture for a Better Future*. Food and Agriculture Organization, Rome.

Friedman, T. (2009) *Hot, Flat and Crowded: Why We Need a Green Revolution – And How It Can Renew America*. Picador Publishing, New York.

Gouse, M. (2013) An evaluation of the gender differentiated impact of genetically modified crop adoption: a pilot study in South Africa – GM maize and gender. Evidence from smallholder farmers in KwaZulu-Natal, South Africa. Project report to the Program for Biosafety Systems, International Food Policy Research Institute.

Gruère, G. and Sengupta, D. (2011) Bt cotton and farmer suicides in India: an evidence-based assessment. *The Journal of Development Studies* 47(2), 316–337. DOI: 10.1080/00220388.2010.492863.

National Academy of Sciences (2016) Genetically engineered crops: experiences and prospects. Available at: https://nas-sites.org/ge-crops/ (accessed 4 February 2019).

# 4 The Politicization of Risk: Is It Safe?

*Sometimes our willingness to consume some type (or piece) of food isn't about whether it's safe to eat, but rather whether one is going to be able to swallow the mouthful that we have taken. Often our brain kicks in before our stomach has a chance to vote, influencing us into thinking there's no way that particular food item is going in this mouth. Foods that frequently fall into this category are liver, tripe, boiled spinach, haggis and, for me, meatloaf (to this day, I remain convinced that meatloaf was primarily developed to poison people in historic times!). Our mental gag reflex kicks in well in advance of seeing or smelling the food, automatically dispositioning us to utterly reject the notion of eating. I've had haggis by the way, and it is good. I also grew up eating liver on the farm and while it wasn't my favourite, it didn't kill me. I can't say I've ever had the opportunity to eat tripe.*

*Most of us get our food prejudices and preferences from our family. My maternal grandfather immigrated to Canada from England as a 17-year-old immediately following World War I. How many 17-year-olds today would undertake a solo, month-long trip, first sailing from England to Halifax in Canada and then spending three weeks on a train to Saskatchewan to settle with relatives they'd never met, to take up farm work? My grandfather then had seven children. By the time I came along, Christmas dinners, following English holiday traditions, included my grandfather, his children and their spouses, various grandchildren and other family relatives, typically with 25–30 people gathered around the very long table.*

*One particular memory is my aunt's 'special' tomato jelly salad that she prepared each year for our family's Christmas dinner. At some point, when I was in my teens, my aunt had discovered a new salad that was based on tomatoes as the main staple of a jelly salad. It was a pretty shade of red, but that was about as far as I was prepared to go in terms of any other assessment. Oddly enough, I was not alone in this form of risk assessment. Once all the food was set on the table, everyone pulled up their chairs and bowls of delicious and steaming food began to circle to table (always passing to the left of course). As this particular salad circled the table, we all politely took a small spoonful, interested to try such a colourful salad. Bad choice! It was terrible. The challenge then became a game of how you could either manage to add a sliver of it to a mouthful of something delicious, without wrecking the taste of that morsel, or manage to dissipate it around your emptying plate, making it look like you had eaten most of it and were so full you could not finish the last bit of tomato jelly salad. This was a significant challenge.*

S.J. Smyth *et al.* (2019). *GM Agriculture and Food Security: Fears and Facts* (S.J. Smyth *et al.*)

*The following Christmas, as we pulled our chairs to the table and began passing the bowls and platters of food, this same brilliant red tomato jelly salad began to circle the table with nary a spoonful taken. It flashed past me and some of my older male cousins and uncles, who were seated in one longer stretch of seats. In fact, it moved so fast, I thought the spoon was going to fall out at one point. Later in the meal, as items began to circle the table a second time to ensure we all got enough to eat, I noticed that one brave soul had indeed taken a spoonful of the tomato salad. Upon a quick survey of people's plates, I noticed a brilliant red blob on the plate of my aunt, who had prepared the salad.*

*My jelly experience taught me that while something may look attractive to the eye, such as the bright red jelly salad, this is not necessarily enough to entice anyone to try it a second time. While there was no legitimate food safety threat from eating this tomato jelly salad, no one was prepared to offend their stomach by eating it. In essence, my family decided that although it was completely safe to eat, we simply were not going to do so.*

## Food Production Options

For millennia, all the food that humans consumed was organic, with no synthetic chemicals or fertilizers used. Farmers in ancient Greece, Rome and Babylon all used manure, wood ash or various natural minerals as a means of supplementing crop yields. Domesticated plants are considerably less efficient at competing for resources than their weedy relatives and so continual efforts were required to ensure that the crops planted would result in sufficient yields to secure food supplies. From our history lessons, we know that crop failures throughout history were frequent and destructive.

As we moved into the 20th century, the use of synthetic chemicals and fertilizers became commonplace, especially in the post-war period. These technologies raised the cost of farm inputs, but the resulting yield increases allowed farmers to be more profitable, justifying the continuing and expanded use of chemicals and fertilizers. Eventually, what has become known as 'conventional farming' became a process of using artificially bred seeds, the distribution of synthetic fertilizers along with the seed and the application of synthetic chemicals to control weeds, insects and plant diseases, as required, throughout the growing season. The combination of seed, chemical and fertilizer technologies was enhanced in the post-war era through the development of more powerful and advanced farm equipment. By the later part of the 20th century, conventional farming accounted for virtually all crop and food production in most developed nations.

Throughout this period of rapid technology adoption, a portion of farmers delayed or resisted change, either due to inadequate resources or due to personal preference. Some of these farmers were small, marginal, non-commercial ventures while others, such as organic producers, saw this as an opportunity. Although production of organic crops and foods accounts for only a small percentage of overall output, these products are popular with some consumers. Some with higher disposable incomes have expressed a preference to pay more for food products that have the same nutrition and safety as other related food products but are produced in specific ways defined by the organic movement. Generally, organic crops are produced using conventionally bred seeds and a range of naturally derived fertilizers, herbicides and insecticides. While the organic industry allows a lengthy list of eligible chemicals (140 chemical compounds are provided in Canada's organic standard) to be used in the production of organic crops, they are careful about how they report this information. The organic industry has spent decades warning the public about the misuse and danger of chemicals in agriculture and certainly don't want to dwell on the fact that chemical use in organic production is both allowed and commonly practised.

The current state of the art in genomic innovation allows plant breeders to genetically modify plants to be more resistant to insects, plant diseases and chemicals. Three main variations of GM crops exist: those that are tolerant to herbicides that are sprayed for weed control, known as herbicide tolerant (HT) crops; those that have insect resistance incorporated into the plant, known as insect resistant (IR) crops; and those that have virus resistance, known as virus resistant (VR) crops. Crops from these innovative breeding technologies were first commercialized in the mid-1990s and have been safely consumed for 23 years.

Trade to improve food security dates back millennia as well. Primitive barter in food products between neighbours and communities evolved into commercial trade that spanned thousands of miles by land and sea. Although much of the early trade involved the rarer and more lucrative items, such as spices, basic commodity trade also existed. Trade, which provided the opportunity to distribute abundant surpluses in one commodity or area to those facing shortages elsewhere, tended to work best in times of peace. Frequent and lengthy wars disrupted and distorted trade flows. With new technologies, food trade has expanded massively. Only five countries are uniquely food self-sufficient, and even they are trade exporters and importers. Technology and trade are inseparable.

Regulation of safety and efficacy is challenged to address both new technologies and diversified and differentiated market needs. When it comes to the regulation of how crops are approved, there is great variability. Agriculture has long used forced breeding, crossings and mutagenesis to develop crops that can be planted as long as they meet basic agronomic requirements, such as disease resistance and yield enhancement, without a rigorous government risk assessment. In some countries, such as Canada for example, the self-regulation is managed by industry groups (comprising breeders, crushers and millers) that approve new varieties before release. There are no federal or regional government risk assessments or approvals done on crop varieties developed conventionally.

GM crops are handled differently. While scientists mostly agree that the technique of gene transfer does not in and of itself confer any higher risk, most citizens and governments are sceptical. Governments everywhere engage in mandatory intensive assessment and evaluation of every new GM introduction, whether it comes about through gene splicing or the more recent gene editing techniques. The rest of this chapter discusses the emergence and management of the array of interlocking and overlapping national and international systems to assess the risk of these crop innovations.

## The Regulation of GM Crops

The regulation of GM crops varies by country. Every country, however, adheres to a common set of basic principles of risk assessment that have evolved over the past 50 years. As discussed in Chapter 3 (this volume), the concept of structured pre-market assessment of products or developments began first in the USA. During the 1980s, the principles underlying the risk analysis framework were articulated and reduced to practice, first in the USA and then transferred to other countries directly or by international normalization though the OECD, the Codex Alimentarius, various standards-setting bodies and ultimately through inclusion in a variety of international environmental and trade agreements.

Risk assessment is fundamentally grounded in science. Scientists and regulators start the process by defining a potential development, agent or product that is suspected of generating risk. Most assessments are forward looking, in that they look at new opportunities, but sometimes they revisit past decisions, for example, many countries are reassessing the risk related to various chemicals approved in the past. Once a suspect is defined, scientists and others will work to determine any potential causal impacts that would generate harm, and then will work to test whether those potential risks are real. At this point, the purpose is to validate the concern, rather than quantify the impacts themselves. For instance, the regulators will look at the biochemical mechanisms at play to determine whether specific compounds can lead to toxic, carcinogenic or other effects. For example, many chemicals may cause mutations, tumours or chemical toxicity. Once the potential for harm is determined, studies are undertaken to determine the likelihood of those harms emerging. This involves dose-response studies, in which a test plant, animal or microbe is exposed in varying doses to the chemical or compound of interest, to demonstrate at what level harm emerges. Living organisms and environments are remarkably hardy and resilient, so the potential for harm does not always necessarily translate into actual harms. Most organisms and environments can sustain some exposure or stressing to potential risk factors without exhibiting any material damage; but usually there is some threshold of safe use or exposure which, once crossed, leads to measurable harm. The challenge of any regulatory system is then to determine where these threshold events are and to decide how much harm is socially acceptable. Science can advise on the cause and effect relationships, but only democratically elected governments can legitimately decide on where to draw the line for what is acceptable. All of these steps are incorporated into what is now called the risk analysis framework, which begins with technical assessment, proceeds to the managerial decision making and ultimately culminates in a range of communications efforts, both to inform the users and consumers of the appropriate use of a risky substance, but also to engage in a two-way dialogue about users' experiences and any unintended impacts that were not anticipated or considered in the formal assessment.

Ultimately, the goal of this system is to manage risk, and not to drive risk to zero. Nothing that is produced for human consumption can be guaranteed to be perfectly safe. Mishandling and misuse of even fundamentally safe foods can lead to harms.

The following overview is based on the framework of government acts and resulting regulations that govern the risk assessment process that is required for the approval of GM crops in Canada, which, to a considerable degree, broadly reflects processes across the globe.

The Canadian governance system for crop agriculture, as for most products and markets, is based on an extensive, horizontally based, blended public–private regulatory system. Risks are managed by various stakeholders depending on the stage of development. Researchers in public and private organizations are responsible for managing any risks in their research programmes as long as the materials remain in isolated conditions (e.g. in laboratories or under glass in a greenhouse). Individuals are, however, extensively governed by internal and external rules. Most researchers are required to comply with the norms and ethics of good research protocols, good laboratory practices and a range of workplace safety policies, all which are established within the institution, by third-parties or by the government to help direct and manage risk. Once a research programme has developed something that they think could be useful and valuable – be it a new variety, new technology or new chemical – they need to begin to test whether it will actually work. Some of that work can be done in laboratories and under glass. When it comes to crops, regulators require evidence that any new variety is genetically stable, which is usually demonstrated by seven generations of breeding true, some of which is done in confined field trials, which are undertaken in selected sites under licence from the government and with strict controls on how they will be managed. To date, there is no evidence that any transgenic food crop that has been managed in laboratories or confined field trials has escaped into the open environment. The few inadvertent releases that have been detected in the rice and maize business have been traced back to firms mishandling materials that were being multiplied as seed in advance of anticipated regulatory approval.

Once a variety has been confirmed as stable and uniform, it is ready to be examined for registration; the formal government regulatory system takes over at this point. When it comes to the production system for crops and food, government has enacted a series of regulatory processes designed to proactively assess the safety implications of commercialization of a new variety. Most regulatory systems, including Canada's, are designed to assess the risks prior to the products being commercialized rather than the post-approval management of any risks and liabilities. The Food and Drugs Act (1985) sets rules for human consumption, the Feeds Act (1983) sets maximum tolerances of nutrients for livestock feed and the Seeds Act (1985) specifies the performance standards for new varieties. These three acts are designed to establish standards for, and manage risks related to, plant agriculture. The Food and Drugs Act establishes risk thresholds for allergenicity, toxicity, metabolization, nutrition and dietary exposure relating to human consumption. The Feeds Act defines the thresholds for the potential risks due to allergenicity, toxicity, digestibility and dietary exposure relating to animal feeding. The Seeds Act requires new cultivars to be uniform, stable and unique; the mandated review of new cultivars by recommending committees considers all of the evidence from the Food and Drugs Act and the Feeds Act, as well as a range of environmental safety considerations, including: the potential of the plant to become a weed or to be invasive in natural habitats; the potential for gene flow to wild relatives; the potential for a plant to become a plant pest; the potential impact of a plant or its gene products on non-target species; and the potential impact on biodiversity. The integration of these three acts into the regulatory framework is designed to identify all potential risk categories and ensure that any new plant variety is benchmarked to existing varieties already determined to be safe for human and animal consumption and in the environment. This process is known as 'substantial equivalence'. While some assert the substantial equivalence standard is flawed, the reality is that it anchors risk decisions on new products in the context of the risks now faced in the market. While in theory it might limit our collective ability to reduce our risks with new innovations, in practice many, if not all, of the new GM cultivars approved have significantly reduced the environmental footprint of food production, with no attendant change in the risks of our diet or for our husbanded animals.

The Canadian system significantly differs from other systems in that it has established and uses the plant-with-novel-trait trigger. Any plant that presents for introduction into the Canadian market with a trait not otherwise in use in Canada would need to be assessed. While other countries narrow their focus to plants that have been modified by transgenic technologies, the Canadian system uses the functional properties of the resulting plant as the trigger. So, it doesn't matter in Canada whether the plant with a novel trait is simply a transfer of a naturally occurring plant from another country or whether it is a product of conventional breeding, mutagenic transformation, transgenic technology or gene editing. As a result, every herbicide tolerant (HT) variety application that the Canadian Food Inspection Agency (CFIA) has received has been treated as a plant with novel trait (PNT), regardless of the technology used to create the HT variety. For canola, that means that the two main transgenic varieties (Roundup Ready™ and LibertyLink™) and the main mutagenic variety (Pursuit™) have all been subject to review. Furthermore, mutagenic varieties of wheat and barley, even without the HT traits, have also been reviewed as PNTs.

Unlike the CFIA, which uses a product trigger, Health Canada defines novel foods as foods resulting from a process not previously used for food, products that do not have a history of safe use as a food or foods that have been modified by genetic manipulation, genetically engineered foods or biotechnology-derived foods. Health Canada assesses the safety of all GM and other novel foods proposed for sale in Canada. Companies are required to submit detailed scientific data for review and approval by Health Canada, before such foods can be introduced as food. Health Canada does not review all foods new to the Canadian market but only those that are deemed novel. The concept of prior safe use as a food was introduced to exclude foods new to the Canadian market which have a history of safe food use in other countries from being the target of a novel food notification. In addition, the concept of 'major change' was introduced into the novel food definition in order to avoid the potential of a minor processing change to trigger a novel food notification. This approach is intended to restrict novel food notifications due to introduction of new processes only to those that are truly new and cause substantial changes in the composition of the food.

While no formal standard or definition for 'novel' exists, Canadian plant breeders use a rule of thumb that if the specific trait they are selecting for expresses at 20–30% higher or lower than conventional varieties, the plant breeder initiates discussions with regulators regarding the applicability of PNT regulations in the specific instance. The PNT regulations apply to all plant varieties having a novel trait, regardless of how they were developed, meaning that the variety could be developed by gene editing, genetic modification, mutagenesis or even old-fashioned pollen-transfer breeding. It is expected that some of the gene editing technologies may create products that are PNTs, while others may create products that are not PNTs.

This rigorous system has worked most effectively, as after 25 years of approving GM crops, there has not been a single environmental or human health risk from a GM crop. Nevertheless, no one in the food system would, or even should, claim that any food is perfectly safe for consumers or the environment. All commercially cultivated plants, animals and microbes alter our natural environment, and virtually everything we eat poses some risk to someone – sometimes because of what is in the food (e.g. allergens, natural toxins or pathogens) and sometimes because of how we handle and prepare the food (e.g. contamination in storage or undercooking). The issue is not whether it is absolutely safe in all contexts, but whether we are identifying and managing the traits that can and do present real dangers to consumers. Our global food system has gone a long way to make foodborne mortality a minor consideration for the vast majority of the global consuming public. The remaining safety concerns are differentially found in developing countries where refrigeration and safe handling practices are less rigorous.

## Food Safety Facts and Fiction

Millions of our ancestors died from the consumption of unsafe food. As recently as 1900, foodborne diseases caused the third highest number of deaths in the USA, behind influenza and tuberculosis. Over time, science-based food safety testing and detection mechanisms have been developed and implemented to improve the quality of food that is consumed, so that in 2018 foodborne diseases in the USA have reduced in incidence and are not in the top 10 causes of death. Of course, no system functions at 100% safety and occasionally consumers die from the food they eat. Unsafe food containing harmful bacteria, viruses, parasites or chemical substances causes more than 200 diseases, ranging from diarrhoea to cancers. The World Health Organization (WHO) reports that an estimated 600 million – about 1 in 12 people in the world – fall ill after eating contaminated food at least once each year and 420,000 die; children under 5 years of age experience 40% of the foodborne ailments. To the best of anyone's ability to determine cause, none of those illnesses or deaths are because of any of the GM foods in the market today.

Risk assessments have become increasingly discerning in their ability to identify additional subsets of the general population that might be at risk from any innovation undergoing regulatory scrutiny. Of course, the cost of preventing more narrowly defined harms for subpopulations rises as other risks are eliminated. The rising costs associated with assessing the potential impact of new food products on ever-smaller subgroupings have grown to the point that in the case of regulating GM food products, an inordinate sum of money has been spent regulating products with no identifiable corresponding increase in either food risks or food safety. Science-based risk assessments have proven that GM foods are safe to consume; there are many risks in other parts of the food system that would be better places to focus attention. In effect, the risk assessment system has been politicized. No matter how refined an assessment might be, those who doubt will never be satisfied. They continue advocating for longer, more detailed assessments. As of 2017, more than 280 institutions, organizations and ministries have undertaken more than 3000 scientific and technical assessments; all concur that GM products present no greater risk than existing crops. Nevertheless, some eNGOs still argue that GM crops and farm chemicals are unregulated or under-regulated and, as a result, should be deemed unsafe. In the following examples, we assess some of the concerns anti-GMO activists have propagated over the last 20 years regarding GM crops and modern agriculture. Given the resilience of the stories in spite of the extensive scientific exploration, one might characterize them as popular myths.

## Farm chemicals cause cancer

The claim that farm chemicals cause cancer is one of the anti-GMO activists' go-to fear stories that are used to sway the public regarding the dangers of agriculture. Most scientists and regulators would not agree, based on the peer-reviewed scientific studies and the extensive regulatory assessments completed around the world. However, activists reject peer-reviewed evidence and discount the safety determinations of national regulatory experts. Modern agriculture that uses mutagenic or GM seeds, synthetic fertilizers and chemicals is demonized, all the while ignoring that all farmers, including organic producers, use chemicals that can be toxic, regardless of whether they are synthetic or natural.

All chemicals present some risk. The challenge is to use them appropriately. While there has been extensive review of the new GM HT varieties and their related chemicals, none of which has identified any particular risk, other farm practices have had less review. Research examining the use of chemicals in organic agriculture has found that the lower efficacy rate of natural chemicals leads organic farmers to spray each field up to seven times to achieve the same level of control as a conventional farmer could achieve from just two applications of synthetic herbicide. This can, and sometimes does, affect the food we eat. An investigative journalist in Canada recently reported that the CFIA tested organic produce from grocery stores and found chemical residues on almost half of the produce tested.

The ongoing fight over the role of glyphosate in our food system exemplifies the challenge. The recent jury decision in the California superior court that found in favour of a school groundskeeper suffering from cancer and against Monsanto, has moved this from inside the news section of most newspapers to the front of the business section. Glyphosate has been used since 1974 in over 160 countries, with over 800 peer-reviewed scientific studies quantifying the safety of glyphosate, including rulings by the WHO and three of its subagencies, the Core Assessment Group, the International Programme on Chemical Safety and the Guidelines for Drinking-water Quality. But in March 2015, the International Agency for Research on Cancer (IARC), another subagency of the WHO, classified glyphosate as 'probably carcinogenic' to humans. IARC appears to have ignored the evidence of the peer-reviewed scientific studies quantifying the safety of glyphosate, basing its classification instead on a small number of dissenting publications. Scientists, regulators and journalists all dug in to find out how such a discrepancy could emerge. Earlier draft versions of sections of the IARC report, drafted by a subset of scientific experts, appeared to be concluding that there is no link between glyphosate and cancer in animals, but the final assessment reversed that finding. Who influenced the change and why it was made remains hotly debated in the scientific community. WHO stands behind their process, but the lack of transparency of the process suggests committees like this are far from perfect arbiters of science.

Experts that review the safe use of chemicals for governments in various countries around the world reviewed the IARC report on glyphosate and within weeks rejected the conclusions that glyphosate is 'probably carcinogenic'. One of the first to categorically reject the IARC classification was the EFSA, followed by national governments in Canada, the USA, Australia, New Zealand and Germany. These leading agriculture-producing countries all use glyphosate as part of crop and food production, having approved and then renewed glyphosate for further use on numerous occasions. Given the length of time in the market, many of these countries had scheduled to undertake reviews of the chemical and any new evidence that has emerged since it was introduced into the market. These reviews included assessments of the potential for glyphosate to cause neurotoxicity, immunotoxicity, chronic toxicity, cancer, reproductive and developmental toxicity (especially links to neurological diseases such as Parkinson's). To date all of the national reviews have led to renewals of the chemical. No evidence of harm was found in any of the reviews.

## Human health and GMO consumption

Anti-GMO activists have worked since the very early days of GM crops to try to show linkages between GM crops and human health issues. Even after 20 years of time and effort to assess each and every introduction, not a single regulatory agency has found any compelling evidence of harm and no compelling peer-reviewed studies have emerged. Nevertheless, activists continue to propagate claims about risks in our food. Scientists trained in immunology and toxicity have reviewed those claims and found a range of errors and flaws in the challenges that in their opinion invalidate the claims put forward by anti-GMO activists.

One of the very early coordinated attacks against GM crops was triggered in 1998 when Dr Arpad Pusztai at the Rowett Research Institute in Scotland issued the results of his studies on the nutritional effects of GM potato. He had been funded by the UK's Scottish Office to assess the impacts of GM potato consumption on human health. The other two groups participating in this project were the Scottish Crop Research Institute and Durham University. In scientific experiments, it is crucial for identical methodologies to be established and followed by various laboratories, allowing for an apples-to-apples comparison of the experiment results. In these experiments, GM potatoes were being fed to rats to study effects on nutrition, intestinal development and immune-system development. The global protocol for an experiment of this type is to establish a control group that receives one treatment and an experimental group that receives the item being tested. Proper procedure is to feed the rats for 90 days. As these experiments proceeded through 1996 and 1997, the three groups met periodically to share results and scientists from both the Scottish Crop Research Institute and Durham University are reported to have expressed significant concerns about the research protocols being utilized by Pusztai. Seemingly, from the very start of the GM potato research, scientific experts were disagreeing with the experimental methodologies used by Dr Pusztai. The experiment that Pusztai was conducting used genetically engineered Desiree Red potatoes that expressed the GNA gene from the snowdrop plant, which produced a lectin protein that had been shown to be toxic to some insects (the specific GM potatoes chosen were neither in the market nor intended for commercial release). He undertook four feeding trials: (i) a control group fed raw potatoes for 10 days; (ii) a group fed raw GM potatoes for 10 days; (iii) a group fed cooked GM potatoes for 10 days; and (iv) a group fed raw and cooked GM potatoes for 110 days.

If the work had proceeded along the normal research path, the results would have been summarized in a journal article and submitted for peer review. Dr Pusztai, however, jumped the gun and went on the Granada Television's current affairs programme, *World in Action*, in June 1998, and reported that the rats suffered significant harm from consuming the GM potatoes. His most controversial statement was that it would be 'very, very unfair to use our fellow citizens as guinea pigs'. His on-air performance ignited a firestorm of controversy and conflict. When his results were peer reviewed they generated even more controversy. His home institution, the Rowett Institute, reviewed his work in October and concluded that his data did not support his conclusions. In early 1999, the Royal Society empanelled six anonymous reviewers to assess the experiments and results. The resulting review in June concluded that Dr Pusztai's experiments were poorly designed, contained uncertainties in the composition of diets, tested too few rats, used incorrect statistical methods and lacked internal consistency. The Lancet waded in to the controversy, denouncing Dr Pusztai, the media and the Royal Society for its review. Shortly afterwards, *The Lancet* sent Dr Pusztai's paper to six reviewers (rather than the normal two) and ended up publishing it as a letter in October 1999, following a split decision among the reviewers (four found it acceptable; one thought it flawed but wanted it published so others could see the data; one was against publication). A subsequent review commented that the only fact that could be taken from this research was that a diet of strictly raw (or cooked) potato was so nutritionally poor that all the rats in the experiment showed effects of starvation.

This research was re-engaged a decade later in France, this time involving an assessment of GM maize consumption and whether this consumption was linked to cancer. In this instance, Dr Eric sSéralini undertook to feed GM maize to rats for a period of two years to determine the long-term effects of consuming a GM food crop. This experiment was undertaken in spite of the fact that risk assessment science conclusively shows that long-term studies provide results that are not different from those of 90-day studies. In medical research, a range of rats and mice are used as test subjects. In this case, Dr Séralini chose a particular strain of rats for his trials that were genetically disposed to develop tumours, at rates in excess of 70% even without any triggering event. A second complication is that he continued the experiment for two years, which is beyond the expected normal lifespan of the average rat. As one might expect, the rats in this study developed cancer, the only question is whether it was due to their diet or just a natural result of old age and a genetic predisposition to cancer. The research was accepted by the *Food and Chemical Toxicology* journal for publication in September 2012, and Dr Séralini engaged in an extensive pre-publication media blitz, so that when it was released it was front-page news. Subsequent review by scientists challenged the study for having inadequate controls and sample sizes that in their view negated the conclusions; there was also controversy about whether there was any dose response, as animals fed less GM maize in their diet seemed to suffer more cancer than others in the study. Ultimately, the journal retracted the study, but it was republished two years later with more data but the same contested conclusions. In response, the EU and France undertook three studies (GRACE, G-TwYST, and GMO90+), to assess the toxicity of herbicide-tolerant (Roundup Ready™) maize, both over the conventional 90-day test period and for longer durations. The research released in 2018 identified no potential risk from the product and reaffirmed that longer trials were not justified scientifically. While this might appear to have resolved the scientific question, many environmental groups (and a few scientists) continue to dispute the whole premise that transgenic foods can be safe under any conditions.

The challenge that we, and we suspect you, have is that when scientists duel about their results, we all get nervous. If the Pusztai and Séralini controversies had stayed inside the academy and on the pages of journals, we would have ultimately been informed of any valid conclusions. But the decisions of Drs Pusztai and Séralini to go public and generate public excitement did no one any good. It simply frightened people in ways that has led to a poorer policy than otherwise would have been the case.

## GMOs and chemicals kill butterflies and pollinators

One of the very earliest scientific criticisms of GM crops was the supposed devastating effect they could have on beneficial insects, such as monarch butterflies or pollinators like bees. Once again, these criticisms were based on hypothetical situations, with no data to substantiate the claims, which have all turned out to be baseless, with no connection to GM crops. There is no doubt agriculture changes the environment for insects. In the first instance, large, relatively homogenous commercial fields offer a great feeding ground for some insects while at the same time disenfranchising others who may not thrive on the plants or other flora and fauna within cultivated fields. Second, most crops have chemicals applied, whether they are GM or not. While chemicals are intended to be toxic to target insects, the goal is to minimize the effect on non-target insects, many which offer beneficial effects in the farming ecosystem. Sorting out which chemicals and farming practices affect which insect has been a challenge. Regardless of where one stands on the safety of GM technologies, one can appreciate the greater insights into the ecosystem effects of crops that materialized once we started to follow their effects. For example, the ability to track pollen from GM canola crops in Canada revealed that bees, the main pollinator for commercial crops, fly up to 30 miles in search of food. Similarly, we learned the likely and practical limits to pollen flow from a range of pollenating crops, as we could differentiate where pollen went through DNA assays.

Within a couple of years of the first commercial production of GM maize, Dr John Losey, a researcher at Cornell University, conducted a laboratory experiment to determine the effects of Bt maize on monarch caterpillars. He fed milkweed leaves that had been dusted with pollen from Bt maize to a population of monarch caterpillars. The logic was that wind-borne maize pollen can settle on the leaves of milkweed plants that monarch caterpillars eat and that often grow in meadows or fields near maize fields. He found that indeed the butterfly caterpillars fed pollen from Bt maize ate less, grew more slowly and suffered higher mortality than the control population. This was not unexpected as the GM maize used was designed to kill European corn borers, which are part of the same *Lepidoptera* family as monarchs.

In many ways, this study should not have excited much attention. It showed the expected toxic response but said nothing about the dose–response relationship. There are two questions that would need to be answered to tell us anything about the appropriateness of the technology. First, would monarch butterflies lay their eggs on plants that would expose their larvae to toxic levels of Bt pollen? Second, and in some ways more important, what other insect regimes exist and what is the relative impact of Bt maize versus the next available alternative?

When Losey's article was published, it triggered significant additional research by other university scientists and 2 years later, a special issue of the *Proceedings of the National Academy of Sciences USA* (*PNAS*) contained a set of these additional studies. A number of issues about the first study were noted. First, the density of pollen sprinkled onto a leaf by hand in a laboratory is vastly different to the density of pollen on a leaf in nature, where wind and rain affect the presence of maize pollen. Second, placing the monarch larvae on the surface of a leaf made a big difference compared with being on the underside of the leaf, where they are actually laid. When larvae hatch, they take days to eat their way through the leaf to the surface, where they would first come in contact with GM maize pollen. This elapsed time allows the larvae's digestive system to mature, arguably to that which should be capable of surviving incidental GM pollen ingestion. Most importantly, though, is that milkweeds are simply not present in GM maize fields. Farmers spray herbicides that kill weeds shortly after the maize emerges from the ground, and milkweed is one of the weeds that are controlled. The work reported in the special issue of *PNAS* reported on all these outstanding issues, concluding that GM maize would have impacts on non-target organisms that are no different than other varieties of maize. Different studies examining the impact of GM maize versus other cropping choices suggest that in fact Bt maize harms beneficial and non-target insects much less than the conventional practice of spraying synthetic chemicals.

Interestingly, research into GM maize has revealed an unexpected but positive benefit in terms of improved food safety. GM maize is designed to be resistant to European corn borer larvae, which otherwise would damage the cob, exposing it to air and increasing the levels of cancer-related mycotoxins in maize. A recent study by an Italian team led by Elisa Pellegrino found that GM maize production has reduced mycotoxins by 30%.

More recently, there has been extensive controversy about the impacts of the agricultural chemicals called neonicotinoids (known as neonics) that are commonly used to coat seeds. Some scientists and studies suggest these chemicals are connected to recent changes in pollinator populations. Pollinators are crucial to crop and food production. Although determining the pollinator population is difficult, one measure is the total number of managed beehives, which has steadily risen since the 1960s. Pollinators encompass all types of insects, including bees, wasps, hornets and thousands of others. There are over 18,000 species of bees, which are the most studied of pollinators. There is limited understanding of wild bee species and the impacts of wild, non-bee pollinators. By 2005, biologists and honey farmers were reporting dramatic declines in beehive colonies, which was coined 'colony collapse disorder'. Almost immediately some civil society groups pointed to farm chemicals as the cause. Although there have been several hundred studies on the relationship between pollinators and changes in their environment that impact their population, such as agricultural chemicals, land-use practices, climate change, pollinator diseases and genetic disorders, very few of these actually provide evidence arising from the natural environment.

As noted in the Losey study, it is problematic to use studies conducted in laboratories to draw substantive conclusions about the natural environment. Studies of chemical impacts in laboratory experiments can demonstrate lethal effects of the chemical on pollinators. These links are virtually impossible to replicate in natural experiments, largely given that concentrations of chemicals are vastly different in outdoor environment conditions than those used in a controlled laboratory experiment. Duplicating a laboratory experiment on pollinators and with pollinators in the wild is exceedingly difficult, with few of these being successfully completed. This is complicated by the fact that the scientific community has no agreed methodology on how to undertake such assessments. A mix of results have been found, but no compelling overall message. Several meta-analyses (essentially a study of studies) have been undertaken, concluding that, when all the variables in play in the experiments are accounted for, there is no measurable relationship between chemical use in agriculture and changes in pollinator populations.

While science has demonstrated that GM crops do not appear to harm butterflies and farm chemicals do not appear to impact bees, this has not stopped some eNGOs from claiming that there is a direct and verifiable connection between GM crops and our butterfly and bee populations. They assert that we cannot wait for further studies – we must ban GM crops and farm chemicals now. The problem is that the alternatives are unambiguously less sustainable, leading to greater use of more toxic chemicals that cause more harm than benefits and affect non-target insects, while at the same time impairing global food security.

## The Future of Farming and Food Safety

Farmers are stewards of the land, the environment and the diversity of crop and food production. Farmers have always adopted technologies and practices that minimize the impacts of food production on the surrounding environment. The fact of food production is that it has an impact on the environment and to suggest that it can be otherwise is both dangerous and misleading. The peer-reviewed evidence to date, more than 3000 studies and the summative judgement of more than 280 regulatory and scientific reviews over the past 25 years have all concluded that development of GM crops and new farm equipment has allowed food production impacts on the environment to decrease substantially since their adoption.

Decades of effort have been invested to ensure that the global risk assessment system for new food products and technologies is as rigorous as is possible. Global standards exist and competent national authorities have the capacity to undertake full and appropriate scientific assessments. Science is at the heart of this system. Science has helped us select and sort through all the possible technologies to deliver a safe and efficacious toolkit of GM crops. But this system is far from perfect. The rising costs and increasing uncertainty of complying with these systems have created barriers that have stalled many potentially safe and efficacious options. This might reassure some consumers and citizens (we are sceptical) but it certainly hurts farming, subsistence consumers, and, we would argue, the environment itself.

While the safety of GM crops and foods has been thoroughly assessed, anti-GMO activists have simultaneously been engaged in efforts to undermine the credibility of the global risk assessment and food safety system for GM foods, with the unintended effect of raising doubts about our entire food system. Realizing that the science was exceedingly effective at dealing with claims of harm, danger and damage resulting from GM crops, many critics have shifted focus and attempted to politicize the system itself. The politicization of risk has led to the discounting of scientific evidence, replacing it with emotion and fear. While there are several thousand published research articles confirming the safety and environmental benefits of GM crops and chemical use, eNGOs have convinced quite a few people that the science is flawed. Consumers in most countries have interpreted the haze from these controversies as signalling a fire to be avoided. In the USA, recent surveys report less than 40% of the general public believes that GM foods are safe to eat. Interestingly, the public in the USA trusts scientists the most when it comes to speaking about food safety, but they are uncertain about what scientists think – even though surveys of scientist members of the American Association for the Advancement of Science in the USA show 88% of scientists believe that GM foods are safe. Clearly, the controversies have muddied the normal translation of scientific opinion, which people say they respect. This has major effects. The EU has mostly rejected the technology, imposing a significant penalty on their own producers, the global market and on those African farmers who are afraid of losing access to the European market. In response to the uncertainties about glyphosate in particular, Sri Lanka banned glyphosate for a period of three years, incurring significant farm income losses and increasing problems managing weeds; ultimately the ban ended as the costs mounted.

At the end of the day, we need food to survive. While some assert that we need a better food production and distribution system, none of their solutions achieve this aim. In today's world we need increasing volumes and quality of food. Our investments in advanced biosciences to improve the productivity, quality and environmental sustainability of our food system have yielded great returns. Throwing them away to follow some whim that defies science will not improve our lot.

## Further Reading

Kiiski, H., Scherer, H.W., Mengel, K., Kluge, G. and Severin, K. (2016) Fertilizers, 1. General. In: *Ullmann's Encyclopedia of Industrial Chemistry*, pp. 1–30. Available at: https://doi.org/10.1002/14356007.a10_323.pub4 (accessed 4 February 2019).

Losey, J.E., Rayor, L.S. and Carter, M.E. (1999) Transgenic pollen harms monarch larvae. *Nature* 399(6733), 214. DOI: 10.1038/20338.

National Academy of Sciences (2016) *Genetically Engineered Crops: Experiences and Prospects*. Available at: https://nas-sites.org/ge-crops/ (accessed 4 February 2019).

World Health Organization (2005) *Modern Food Biotechnology, Human Health and Development: an Evidence-Based Study*. Available at: http://www.who.int/foodsafety/publications/biotech/biotech_en.pdf (accessed 4 February 2019).

# 5 Industrial Agriculture: Who Is the Real Winner?

*In this world of heightened awareness and concern about the agglomeration and concentration of wealth by the one-percenters, or even the top 1% of the one-percenters, a big fear regarding technological change in agriculture is that all the benefits flow to the handful of multinational agro-chemical conglomerates or to the branded-food processors or retailers, leaving nothing but risk and uncertainty for farmers or consumers. On the face of it, the willingness of large companies such as DuPont, Bayer and BASF to pay hundreds of billions of dollars to merge or acquire assets, operations and products that they think will add value to their operations suggests there may be some truth to that fear.*

*While the myth of the robber baron taking all the gains remains vibrant, the evidence shows otherwise. At one level, the fact that more than 18 million farmers in about 30 countries have willingly (sometimes enthusiastically) tried and used the technologies, suggests that they must be getting returns above those they are replacing. In those countries without any regulatory, technical or commercial barriers to adoption, the uptake and use of GM seeds has been both fast and significant. Some estimates put the adoption rate in Brazil, Canada and the USA at more than 90% in those crops with competitive GM seeds. The question remains, however, about who shares the gains.*

*We can report that a large number and variety of farm-level, industrial and sectoral studies and analyses have identified an unexpectedly large and diffuse set of positive impacts. Benefits are larger than many people think and are shared much more widely and equitably than feared, while costs so far have been limited to a few sections of the agro-industrial space and, as of now, are more than covered by the gains.*

## Figuring Out Who Wins and Who Loses

Two key issues determine the commercial sustainability of any innovation. First, there must be adequate returns to justify the investments in research and costs of adopting it. Second, the returns must be distributed in such a way as to sustain the participation of all the necessary actors in the system, in our case farmers, seed companies, multinational biotechnology companies,

S.J. Smyth *et al.* (2019). *GM Agriculture and Food Security: Fears and Facts*
(S.J. Smyth *et al.*)

branded-food processors, the wholesale and retail system, public research agencies and, last but not least, consumers.

There are multiple layers of analysis one might consider in determining the impact of an innovation. Most people would simply look to see where the revenues or profits accrue from the production and sale of the innovation itself. That would tend to focus on the few multinational firms that have developed and now produce and sell the innovations. Clearly, they are core to any discussion as they invest billions in the development of products and generate multiple times the value of their investments in gross receipts by selling their successful innovations. Undoubtedly these firms are gaining from their efforts. If we look more holistically at the past, present and future costs, benefits and impacts of the key innovations on both the innovating company's performance and on those that it directly competes with, we see a more nuanced story of a few winners, some unambiguous losers and a range of others who are affected. The willingness of firms inside and outside the sector to acquire or merge, suggests the value proposition is real and ongoing.

That, however, is just the tip of the iceberg that is the economic effects of a successful venture. Economists estimate the returns to research in a somewhat different way. In addition to the returns to the leading companies, economists investigate who else might gain or lose from an innovation, either directly or indirectly. One estimate is that more than US$186 billion was generated over the first 22 years of the use of GM seed technology. Who gets how much and where it is realized depends on many factors. The list of potential winners and losers gets increasingly long and the impacts widely distributed for most of the technological changes that we are interested in.

We cannot stop with only examining the biotech firms. Their earnings and profits come from the sales they make to farmers. In the case of GM crops, that is a yearly estimate of 18 million farmers in about 30 countries. Serving those farmers involves a highly diverse and variously structured input industry. Ultimately, however, farmers and the supply chain are only intermediaries in the value system. Almost every person in the world is potentially affected by the technology, either as a direct consumer of one of the crops with a GM trait or as a consumer of a competing product.

Who wins or loses from the technology, and how much they win or lose, depends on the costs of bringing the product to the market, the nature of the product itself, how competitive the various stages in the agro-industrial supply chain are, the regulatory system, and the individual and collective preferences of consumers. Much assumption, data collection, modelling and analysis is needed to provide the full picture.

In the first instance, although a technology may generate positive cash flow and profit for the innovator once it reaches the market, this does not necessarily mean it is an absolute win. Firms often invest in research that fails to produce a working technology, and, even if the technology does work, sometimes the sunk costs of reducing to practice, scaling up and complying with regulatory requirements are larger than the net returns from the market. Many firms looking back at past decisions can see the flaws in their choices.

Even if the firm that successfully introduces a new technology or product unambiguously wins, however, the industry or sector as a whole may not. In the biotech industry, for example, there is a limit to the area farmers globally are willing to plant to any crop at any point in time, so if one firm gains market share due to a new and better offering, it is undoubtedly taking away some revenue from other firms. Furthermore, some technological changes are so unambiguously an improvement that previous technologies are no longer competitive. This process of 'creative destruction' sometimes causes major changes in the net worth of firms and sectors. Consider all the buggy and wagon makers, stables, animal feed providers and manure collectors negatively impacted with the adoption of automobiles.

The functional properties of a new technology, product or service will often define the scale and scope of the benefits and costs of its use. Some innovations simply make a production system more efficient, which, all other things being equal, increases the supply at any given price. Other innovations change the end-use properties of goods or a service, with a corresponding revaluation of what it is worth to the consumer. Some innovations make a product more attractive, which, all other things being equal, would mean people might be willing to pay more for it; conversely, some innovations make products less valuable, which might cause consumers to discount them.

The structure of the agro-industrial supply chain matters a great deal. If all the segments of a supply chain were made up of small firms that competed solely on price, with every firm having access to the same technologies, then we might simply ignore that part of the chain, because any productivity gains would translate, through competitive action, into lower prices for consumers. Few segments of the supply chain, however, are perfectly competitive. As we shall see, many segments of the chain are highly concentrated, which can work to change both the size of the market (i.e. firms may restrict supply to exploit their market power and generate higher profits) and the distribution of the gains among different actors in the market (i.e. farmers and consumers might be willing to pay more for competing choices, leaving them a smaller net benefit from using the technology).

Regulators and the regulatory system, as gatekeepers to the commercial market, have major impacts on the scale, scope and location of the costs and benefits. First, the size and nature of the hurdles for gaining market approval, and thus access, can influence what is created where and by whom. Both the scale and duration of regulatory review can work to delay commercialization, which works to raise the capital cost of innovation and lowers the returns. There is also extensive evidence in both agricultural biotechnology (agbiotech) and other highly regulated industries that higher regulatory hurdles lead to corporate concentration and a bias towards larger and more integrated firms, which can exacerbate the anti-competitive impacts of corporate structures. Another issue is that not all regulators assess and judge these products the same way. The fact that only about 30 countries have approved one or more GM crop means that the scope of the technology is limited, which means that most farmers in the world are unable to participate in testing and using the technology. Similarly, at the other end of the supply chain, consumers may access GM foods variably – some regulators allow marketing of new products with little incremental costs, while others require proactive disclosure of contents and extensive measures to preserve separation between GM and GM-free supply chains (this is called coexistence in the policy debates). Thus, although neither a buyer nor a seller of GM crops, government agencies can have a major influence on the impacts of GM crops.

Finally, individual and collective preferences of consumers vary widely. We are far from completely rational, but we generally buy what we need (or want) at prices we are willing to tolerate. In those cases where GM crops simply change productivity, consumers who are indifferent to the technology, gain from the reduction in absolute or relative prices, while those averse to consuming foods with GM ingredients may either gain (if the overall drop in cost for the GM product drives down the price of substitutes) or lose (if the costs of substitutes is higher). Interestingly, consumers do not even need to consume GM crops to gain or lose. If the technology changes the balance of production to either drive down overall costs or alternatively pull farm area away from non-GM uses, then lower or higher prices respectively might cause consumers to gain or lose accordingly. In contrast, if the GM trait introduces a new quality attribute, then those consumers who value it more highly might gain a bit, but so will the rest of the supply chain. The net impact and distribution may well be ambiguous.

Of course, simple analyses can misrepresent or distort our impressions of reality. The main challenge is that most analyses are partial in nature. Firms look at the returns on investment for specific research projects and applications, ignoring the impacts on others in their sector and beyond. Governments are usually only interested in what a technology may do in their jurisdiction, yet we know that exports of food surpluses to foreign markets can account for a large share of the benefits of any change; if the government of an advanced industrial economy also cares about global food security, then those impacts should be included. Farmers and consumers also make choices that have knock-on effects on others.

## The Need for a Checklist to Assess Industrial Structure

One of the key factors that influences how the returns of any innovation might be distributed is the structure of the industry. Three phases of industrial change stand out: the emergence, specialization and consolidation of the global GM seed business before 2000; the genomics revolution and licensing agreements among the leading companies in 2000–2015; and, most recently, the emergence of new biotechnologies and the consolidation of the multinationals.

The independent seeds business was the first domino to fall in this biotechnology story. As the multinational chemical industry began to focus on agbiotech, a first order of business was to create specific agri-food ventures and to secure access to the market through purchase of breeding programmes and seed businesses. Five big ventures emerged. In 1999, Pioneer Hi-Bred, the oldest maize breeder, merged with the chemical giant DuPont. Syngenta Global combined the agro-chemical interests of Geigy, Sandoz, Ciba and ICI, merged with Astra, Zeneca, Novartis and Stauffer and then acquired a host of seed brands, including Garst, Funk, Rogers, Northrup– King, Zeneca, Golden Harvest and Advanta BV, a range of seed companies in China and flower companies in Europe. Monsanto restructured in 2000 as a pure agri-food operation, along the way spending billions to buy Holden's Foundations Seeds, Asgrow Agronomics, DeKalb Genetics, Seminis Vegetables, Kruger Seeds and a host of other seed companies in the USA and the EU. Dow AgroSciences, formed in 1950 as the agricultural unit of Dow Chemical Company, acquired Mycogen and Cargill Hybrid Seeds. Aventis CropScience emerged in 1999 out of the merger of Hoechst Schering AgrEvo GmbH and Rhône-Poulenc. Bayer CropScience and BASF both consolidated a range of smaller players to also become major players in the early years.

Although seeds were important, the firms quickly realized that genetic refinements and new technologies offered even more opportunity for market power. Almost from the beginning, the firms jockeyed to control the new genetic tools. Major patent infringement cases, which delivered multi-million-dollar settlements, focused firms' attention on the reality that few if any of the major players had full and complete freedom to operate – each of them owned and controlled only a part of the toolkit needed to efficiently and effectively develop new crop products. In effect, the industry faced hold up without some compromises. Meanwhile the emergence of genomics and automated gene sequencing created a race to sequence and patent almost everything. At one point, the big multinational enterprises (MNEs) and almost everyone else were rapidly sequencing parts of most plant, animal and microbial genomes and filing patent applications with little knowledge or sense of what the genetic sequences did. Over time, firms realized that other strategies offered more opportunity. Over the past decade, this has led to extensive licensing of technologies among the majors, to enable competition among the few but exclude competition from startups and new entrants. Cross-licensing usually involves two or more firms or research organizations providing each other research (and sometimes commercialization) licences that allow them to use any or all of the technologies owned by any of the agreeing organizations. This is often cheaper and more predictable than trying to negotiate terms for each individual technology.

Most recently, the industry has moved to consolidate into three global firms. Dow and DuPont merged, Bayer bought Monsanto and ChemChina acquired Syngenta. The US$130 billion merger between Dow and DuPont has resulted in the creation of three companies: Dow will focus on performance materials and coatings, packaging and specialty plastics; DuPont will specialize in electronics and imaging, industrial biosciences, safety and construction, transportation and advanced polymers; and a new venture, Corteva Agriscience, will focus on seeds and crop protection. The merger between Bayer and Monsanto resulted in Bayer having to divest its Invigor canola, juncea and wheat breeding business, which was subsequently purchased for CAN$11 billion by BASF. An estimated 4500 Bayer employees accompanied this transition to BASF.

Before these mergers, the market was already highly consolidated, both horizontally and vertically. The six largest companies controlled 75% of the global agro-chemical market, 63% of the global commercial seed market and undertook more than 75% of global private-sector agricultural research on seeds and pesticides. The mergers reduced that to four companies. More importantly, these mergers have increased horizontal control: the Dow/DuPont and Monsanto/Bayer duopoly will control 75% of the US market for maize seeds, 65% of the market for soybeans and more than half of the market for crop chemicals, while ChemChina/Syngenta, with about 8% of the global seed market, will serve 25% of the agro-chemical market; BASF controls much of the rest.

In spite of all this movement in the biotechnology and seeds industry, some assert we are focused on the wrong end of the supply chain. Retailers – such as Walmart and Carrefour – and branded-food producers – such as Kraft – have significant market power as they are the primary point of contact where consumers exert their choice and variably value new foods. That relationship provides significant market power and control over the nature of what is consumed, how it is supplied and where the value lands.

## Impacts of GM Crops

There are multiple ways to tell the story. At the highest level, economists have estimated that over the first 22 years of cultivation, up to 18 million farmers in some years have cumulatively planted a total of 2.34 billion ha of biotech crops in more than 30 countries. GM soybeans account for 48% of all the acreage, maize for 30%, cotton for 16% and canola for 6%. Only trace amounts of other GM crops were cultivated. Two traits, insect resistance and herbicide tolerance, alone or in what is called a 'stacked' combination, are in almost all the acreage. Eleven 'mega-adopting' countries, defined as those planting more than 1 million ha, accounted for 99% of the effort in 2017: USA, Brazil, Argentina, Canada, India, Paraguay, Pakistan, China, South Africa, Bolivia and Uruguay. In the first instance, one can assume that the immediate benefits for producers, whatever they may be, are concentrated in the hands of those producing the abovementioned crops in these mega-adopting countries.

Overall, since 1995, the cumulative net gain from the introduction of GM seed is estimated to be about US$190 billion. The annual production of about 190 million ha of GM crops in 2017 is estimated to have generated about US$6.3 billion of gross technology fees and charges for the owners of the technology and US$15.4 billion in higher incomes for adopting farmers. A raft of studies shows that in aggregate the use of GM seeds has raised average yields by about 22% and farm incomes by 68%. Although these are important drivers for the story, they are just a part of the impacts of the technologies.

One way to illustrate the nuances of the impacts is to review a few examples of the various beneficiaries of several products at different points in the supply chain around the world.

The best place to start is with HT soybeans, one of the earliest GM crops that now accounts for almost half of the soybean output. Some half-jokingly called Roundup Ready™ soybean 'soybeans for dummies' due to the simplicity and ease of adopting it into the crop rotation. Although 10 countries have planted GM soy, most of the activity has been in the USA, Argentina, Brazil, Paraguay, Canada, Bolivia and Uruguay. Adoption rates have risen to above 90% in most of the adopting countries. In the first instance, the more than 1 billion hectares planted to GM soybeans has generated strong benefits for adopters. Those soy farmers in countries that allowed cultivation of GM crops gained handsomely; the sector captured more than two-thirds of the net benefit of the new technology, shared between producers, the innovators and others in the supply chain. Not only are traditional commodity farmers benefiting, but GM crops are expanding the area cultivated for some crops. For example, in Canada, Ontario and Quebec are traditional producers of soybeans but in 2013 the western province of Manitoba passed Quebec as the second largest soybean region in Canada; Manitoba in 2018 produced twice as much soybean as Quebec. The USA captured about 45% of the net benefit, with Argentina and Brazil jointly sharing about 47% of the gains. Their overall success, however, worked to raise global soybean yields and lower global soybean prices, all other things being equal, which worked to redistribute about one-third of the benefits to consumers around the world. This also put pressure on non-adopters; any soy producer that did not adopt would face more competition from lower-priced soy, with the result that they lost farm income. Farmers in the EU, India and China were all net losers in the first decade of adoption due to their inability to access the technology. Interestingly, the rise in the relative competitiveness of soybeans had significant impacts on other markets. In the first instance, substitutes for soybean oil lost out: canola, sunflower and palm oil producers around the world were worse off (although canola growers in Canada and the USA tended to remain competitive as they had access to equally good GM traits). The rise in the relative competitiveness of soybean also worked to draw cropland away from other less competitive crops, such as wheat, so much so that global supply of those crops shrank at the margin and prices were higher than otherwise.

The maize story is much the same in terms of aggregate impacts, although the USA gained about two-thirds of the net producer benefits. The story offers some other interesting insights, as maize is the main crop targeted on farmers in southern and sub-Saharan Africa, where we can see evidence of the impact on small landholders. Some studies calculated that hand weeding in Africa requires 150–276 h/ha. Those farmers, often women, who planted stacked-HT and IR varieties were able to sharply reduce weeding effort and time, which was found to positively contribute to children in farm families attending school for more of the year and for other family members to pursue off-farm employment that generated higher wages. The net effect is that income per hectare rose sharply, due to a mix of reduced losses to pests, lower labour costs and higher yields, and overall household incomes soared.

Cotton, although not a major food crop, is a key part of the agbiotech industry. Cotton traditionally is attacked by many pests that have been hard to control. Many producers have used highly toxic chemicals to knock out pests, but some have not had that opportunity. As a result, the impacts of IR cotton have been varied. In the USA and other cotton producing countries that have had the resources to use chemicals, the net effect of the new technology has been to lower pesticide use; farmers share that saving with the GM developers, who levy a royalty that recoups about half of the savings. Of course, the net gains depend a lot on insect pressure. In those years with few pests, farmers may actually pay a little for the security of insect resistance, much like when they buy crop insurance and do not experience losses. In years with significant pest pressures, they are unambiguous winners. Other markets gain in different ways. In China, for instance, farmers experienced fewer yield gains because pest damage on conventional cotton was previously controlled by heavy pesticide use. China has been able to significantly reduce its use of chemical pesticides on cotton, with important environmental and farmer health benefits. The crop was also judged to be decidedly good for smaller and poorer farmers in developing nations. The smallest farms (less than 0.5 ha) experienced the largest yield gains while mid-size farmers (0.5–1.0 ha) had the largest reductions in total costs due to lower pesticide use. In terms of net income, the gains for the two smallest farm-size categories were more than twice those for the largest farms (over 1.0 ha). The Indian experience with Bt cotton was different again. Unlike in China, introduction of Bt cotton was very controversial. Some critics charge that Bt adopters were worse off than non-Bt growers. The main issue was that in India few farmers were able to afford to control pests before IR traits were introduced. As a result, yields were depressed. With Bt cotton, some farmers found that yields jumped significantly (up to 80%) but incomes were not as vibrant, as the cost of seeds more than doubled. The net impacts were quite variable, with farmers in some states reporting strongly positive results and some reporting poor performance. Some attributed the variability of impacts to the lack of locally adapted cultivars and poorly controlled seed quality in some states. In some cases, seed merchants simply sold conventional cotton as GM cotton, defrauding their farmer customers.

GM canola is the crop we know best, as it was both developed and adopted first in Canada and is a major Canadian export. Our studies of the economic impacts of GM canola showed similar results to those found in other crops. Our results, however, reveal some nuances. Canola is the third most important edible oil crop in the world behind maize and palm. In the mid-1990s, transgenic, HT varieties were introduced and since then a range of other GM traits have been introduced, including hybridization and a number of novel oil profiles developed for the differentiated food and industrial markets. What makes this story so interesting is that the sector has sustained competition in the seeds trade, with three large companies competing for a share of the growing seed market: Bayer, now the owner of Monsanto's

Roundup Ready™ varieties; BASF, the current owner of LibertyLink™ varieties; and Corteva Agriscience, the agriculture division of Dow DuPont and owner of Clearfield™ varieties. While the three breeding systems control much of the market each year, the specific market share for any of the three technologies varies each year depending on the weed, pest and disease pressures and the competitiveness of the underlying germplasm in the available varieties. The average half-life of a commercial canola variety is down to below 3 years, due to significant sustained increases in yield. A second aspect of this story is that we can now see a change in the distribution of benefits between the MNE innovators, farmers and consumers. In the early years, most of the net benefits went to the MNEs who owned the technology, but as choices expanded and the market grew, farmers gained a larger share and consumers began to see modest but widening declines in the relative price of canola compared against other oilseeds. Third, we were able to construct an industrial view of the returns on investment in GM canola. Overall, we estimated that the firms with successful traits took more than 6 years of commercial growth to recover the net present value of the capital costs of their direct investments in GM canola traits. The overall research in the crop globally is being repaid, but the increased private investment in the 1990–2005 period drove down the overall return on investment to about the cost of capital, suggesting that the 'races' to invent were not without some wasted effort. Farm-level studies of impact also show a number of interesting effects not observed in other crops. First, farmers reported that one of the main economic gains from HT canola was the drop in dockage – historically farmers faced an average discount of 6% for the presence of off-types and weed seeds in their deliveries. Second, farmers discovered that the use of HT canola in some fields infested with hard-to-manage perennial weeds, such as Scotch thistle, were now coming out clean enough to plant to non-HT crops (e.g. wheat) in succeeding years; they reported significant economic gains as a result. Third, farmers adapted their seeding practices, moving canola up to the front of the seeding season, now that they had effective weed management; as long as canola was weedy, it was often the last to be planted, which both limited its yield and narrowed the range where it could effectively grow. Moving canola up in the seeding season triggered a doubling in the area seeded to canola, pushing out both land left fallow in the rotation and less competitive crops, such as wheat. Finally, over time it became apparent that HT canola was especially profitable for farmers using min-till or zero-till technologies (such as air seeders), which worked to incentivize western Canadian farmers to rapidly adopt this new farm management approach.

While most of the studies of the economic impact of GM crops have examined the top four crops, there is at least some evidence that other crops have equally interesting effects. Researchers and the industry argue that without papaya ringspot virus-resistant varieties, the papaya industry would have vanished from Hawaii. Others note HT wheat, HT flax, virus resistant squash and viral and IR potatoes and tomatoes were important innovations to their part of the industry but were shelved or withdrawn due to producer or consumer resistance.

What about those producers and consumers who passionately want to avoid using or consuming any GM traits? The non-GMO community, particularly organic producers and consumers, have aggressively asserted that the mere existence of GM traits in competing crops damages their interests. Many organic producers and consumers campaign against the existence of GM crops in the supply chain or market, arguing that it restricts their capacity to make an unconstrained set of choices, or at a minimum increases the costs of those choices. This has been at the core of the European policy to secure coexistence, specifically managing differentiation between GM and non-GM products. Although countries have used a mix of regulation and labelling rules, so far there is only limited access to both GM and GM-free produce in the same markets. The organic industry has moved to take advantage of that gap in the market. The evidence of any aggregate harm to the organic sector is unclear. Admittedly, while some food groups, especially processed foods using soybean, maize meal or oil, or canola oil, are likely to be at risk of some mingling of GM and non-GM varieties, there are usually available substitutes that can meet even the most exacting demands. Organic produce, foods or ingredients procured from countries without any GM crops are usually able to meet most needs, often with some extra profit for the new source of supply. Interestingly, there is some suggestion in the literature that the organic industry, and perhaps even avid consumers of organic foods, may actually have gained from the introduction of GM crops. The logic is that quality-attributes such as organic certification tend to generate both benefits for consumers and higher demand for farmers. Therefore, it is not inconceivable that many marginal producers who were unable to enter the organic market and consumers who previously were forced to source their foods outside the organic sector, now have the opportunity to engage in mutually beneficial exchange.

In conclusion, we can confidently say that for those GM crops that have been able to enter the market, adopting farmers, most consumers, the producing and exporting countries and the multinational life science companies have all gained. In addition, those not directly engaged in producing or consuming these products have for the most part benefited from generally lower-priced food baskets and overall economic prosperity. While there are some individuals and groups who may perceive that they are made materially worse off by this development, a full review of their circumstances would be likely to show a more nuanced situation, whereby some of the harms assigned to GM crops are actually more appropriately pinned on the broader socio-economic system.

## Trends and Opportunities

The picture we have painted here sounds almost too good to be true. The advent of biotechnology in crop agriculture would appear to have been enthusiastically received by many producers, consumers and countries, while farmers, consumers and the investors mostly gain and, as discussed in other chapters, there are significant gains for the global environment. So what is the problem? The main sticking point about the future of crop biotechnology relates to where it will go from here. When the agro-chemical industry first turned its focus to crop biotechnology, many representatives of the industry spoke with enthusiasm about their 20/20/20 vision: that is, they expected to quickly roll out as many as 20 different transgenic traits in as many as 20 different crops, from large area to niche, in more than 20 countries. Instead, today virtually all of the GM area globally is limited to applying two traits, albeit stacked and somewhat differentiated, to four large-area crops in about a dozen traditional agri-food producing and exporting countries. In spite of the enthusiastic and optimistic forecasts of where GM crops might go next, all we have really seen is infill work to address this narrow range of applications and uses.

The question is whether this is it? Is there no further use for this technology? GM crops are the focus of an increasingly narrow group of global life science companies. Many others have moved on to use both conventional breeding techniques and a host of new breeding technologies to do what transgenics could do, but they hope with less conflict and fewer costs. The MNEs at the heart of this industry are profitable, but their potential growth is limited. Adoption is virtually complete in those countries that allow cultivation. There is still a significant portion of the potential global market not served, particularly food-insecure nations. One of our recent studies showed that the forgone revenue for the global life science companies, alone, totalled in the billions, not including any other crops or traits than those already in the market. If an average portion of those revenues were invested in new and different GM research goals, we might expect on average two major new trait events annually that could overcome a variety of the long-standing biotic and abiotic stresses and nutritional concerns in our existing food system; the forgone opportunity is substantial.

## Further Reading

Gaisford, J.D., Hobbs, J.E., Kerr, W.A., Perdikis, N. and Plunkett, M.D. (2001) *The Economics of Biotechnology*. Edward Elgar, Cheltenham, UK.

ISAAA (2018) Biotech crop adoption surges as economic benefits accumulate in 22 years. ISAAA brief 53-2017: Executive summary. Available at: http://www.isaaa.org/resources/publications/briefs/53/executivesummary/default.asp (accessed 5 February 2019).

Kalaitzandonakes, N., Phillips, P., Smyth, S. and Wesseler, J. (eds) (2016) *The Coexistence of Genetically Modified, Organic and Conventional Foods: Government Policies and Market Practices*. Springer, New York.

Phillips, P. and Khachatourians, G.G. (2001) *The Biotechnology Revolution in Global Agriculture: Invention, Innovation and Investment in the Canola Sector*. CAB International, Wallingford, UK.

Smyth, S., Castle, D. and Phillips, P. (2014) *Handbook on Agriculture, Biotechnology and Development*. Edward Elgar, Cheltenham, UK.

# 6 The Nature of Natural: Whose Preferences Should Dominate?

*While the ability to produce our own food is an attraction, urban living doesn't provide the scale to be able to do this in any meaningful way. Sure, it is nice to be able to have a few fresh tomatoes in the summer from a balcony pot or a backyard plot, but this means of production is capable of feeding most of us for a few weeks at most. To be able to make a serious dent in the groceries one purchases at the local grocery store requires 5 months of part-time labour.*

*In the 1960s and 1970s, the road infrastructure in most rural areas in western Canada was nowhere near what it is today. In winters with above-average snowfall, it was not uncommon to go for stretches of 3 or 4 days unable to get off your farm, due to the one road leading to town being completely blocked with snow. People would have to wait until the municipal grader was able to open the road, but some winters even this became a challenge and so many municipalities started snowplough clubs, whereby they would pay a local farmer who owned a tractor-mounted snow blower to drive around and open roads after a bad storm. Graders only have a big, heavy V-plough on the front, meaning the snow can only be ridged to the sides of the roads, and once the ridge is 10–12 ft high, it gets very difficult to ridge it any higher. When the gap between these ridges blows in with new snow, the road is blocked and the only way to open it is with a snow blower that is capable of blowing the snow above the existing ridge.*

*The uncertainty about being able to get to a grocery store on a regular basis meant that farm families themselves prepared the vast majority of the food that they consumed. For those who grow a few vegetables on a balcony or in their backyard, this may sound quite quaint or idyllic, but it was a summer-long, daily investment of time and effort. Starting with planting the wide variety of vegetables required, there was the daily weeding and the regular need to ensure that everything got the required amount of water, as nature is frequently rather poor at providing this on a consistent basis.*

*Gallons of peas would be picked, shucked, washed, blanched, bagged in meal-sized portions and then placed into a large deep-freeze until brought out for a meal in the coming winter. The same process was followed with green and yellow beans, whereas maize was more laborious as all the kernels had to be cut off the cob, prior to being placed into bags. Dozens of jars of pickles were made later in the summer, once the cucumbers reached a consumable size. Onions, carrots, turnips and potatoes*


S.J. Smyth *et al.* (2019). *GM Agriculture and Food Security: Fears and Facts* (S.J. Smyth *et al.*)

*all had to be dug from the ground and stored in a root cellar in the basement of the farmhouse, as it was cool enough there to keep them from spoiling.*

*The tomatoes and fruits required the most effort to can for the winter. Tomatoes had to be boiled for a few minutes so that their skins could be peeled, then fitted into quart sealers, which were then boiled in a large canner for up to an hour, thus ensuring the jars had tightly sealed. Fruits such as cherries, pears, peaches and plums would have to be cut up and pitted, while crab apples could just be placed into the jar. A sugar-based syrup was then poured over the fruit in the jars, which were then boiled until sealed. Apples would have to be cut up, cored and then frozen in small bags. All of this required a great deal of time and effort to ensure that there would be food available for the long, cold winter months.*

## The Safety Assessment of Food

Food is both the most ubiquitous and arguably most misunderstood part of our lives. At its most basic, food is any substance consumed to provide energy, maintain life, or stimulate growth and cellular renewal. The human diet involves a mix of plants and/or animals that contain and deliver carbohydrates, fats, proteins, vitamins or minerals. At one level, everything we ingest is indistinguishable by our bodies – it simply breaks down into one of the many nutritional elements that we need to survive. But as humans, we like to reframe our reality in our own image. A quick search of the Internet reveals that we have defined a range of food types we are concerned about, including: whole foods; processed foods; health or healthy foods; functional foods; fresh or frozen foods; diet and negative-calorie foods; kosher foods; halal foods; organic foods; peasant and traditional foods; seasonal foods; medicinal foods; and natural foods.

All the foods we eat contain chemicals and most are preserved or processed with some added chemicals. In the vignette above, home canning of tomatoes requires a tablespoon or two of lemon juice per quart to ensure the safe acidity of the preserved tomatoes. This would probably not raise concerns with the vast majority of consumers but lemon juice is citric acid and if one were to indicate that a tablespoon of citric acid had been added to a quart of tomato preserves, some consumers would express aversion to consuming them due to the presence of this chemical. At times, it seems as though part of society has an aversion to chemicals in food and the environment, regardless of the nature of those chemicals. Chemicals are frequently viewed as dangerous and to be avoided at all cost, while people forget that one of the most basic chemicals is a daily requirement to living – water.

Herbs and spices have a long history of use in food preparation and preservation. While the use of both herbs and spices for their medicinal properties dates back to 2700 BCE in China and 1500 BCE in ancient Egypt, their use in food dates back almost as far. While herbs and spices certainly enhanced the flavours of the food they were included with, they also served to mask the unpleasant smells and tastes of meats that had begun to rot prior to consumption. Trade in spices was one of the most valuable in the ancient world, with supplies moving from Asia and India to Europe, the Middle East and northern Africa.

When it comes to the use of chemicals to aid in the preservation of foods, we can see a variety of applications; frozen vegetables require less processing but had to wait in most parts of the world for the development of refrigeration and the cold chain. Peas, beans or maize do not require chemicals to enhance their storage, but when it comes to processing, chemicals contribute a great deal to the safe future consumption of these foods. Chemicals are common in fresh fruits and vegetables. Leafy greens can be a source of unwelcome bacteria that have a nasty habit of making us sick after consumption, such as *Escherichia coli* and *Salmonella* spp., therefore many companies that sell leafy greens, frequently as pre-packaged salads, use chemicals to reduce the risk of bacterial infection. When the company harvests a leafy green, it is put through a water wash that includes a 0.5% concentration of chlorine, which is enough to kill any surface bacteria that may exist on the leaves. In fruits, fresh apples are sprayed upon harvest with a thin coat of wax that preserves the moisture within the apple during transportation and while on store shelves. This non-toxic coat of wax can be washed off by a quick rinse under the tap prior to consumption. The use of chemicals in the preservation of food is designed to increase the safety of the food products available for purchase and to reduce the rate of spoilage in food products, which helps keep food prices lower than they otherwise would be.

Many of the chemicals added to foods are intended to increase the health attributes of the food product, such as is the case with foods that have increased levels of omega-3. Omega-3 is the marketing name of three fatty acids that have known and documented health benefits: eicosapentaenoic acid (EPA), docosahexaenoic acid (DHA), and α-linolenic acid (ALA). EPA and DHA are primarily found in specific species of fish, with ALA present in plant sources, including nuts and seeds, such as flax. While nutritional supplements are a common way to increase one's intake of these healthy acids, numerous whole- and processed-food products have enhanced levels, contributing to a healthy diet.

Chemical variations of sugars, such as sucrose and fructose, and salts, such as sodium, are also used to enhance the taste of the foods we consume. Historically, salt was such a valuable commodity that wages were commonly paid using salt and the word salary is derived from the Latin word for salt. Other chemicals that are routinely added to foods are vitamins. Vitamin D is added to virtually all whole milk products on the market today in most developed countries. Vitamin B, iron and other essential minerals are added to most bread products to replace the nutrients that are lost when whole grains are milled into white flour. Vitamin C is added to many juices, while meat-processing firms are required to add a form of vitamin C (isoascorbic acid or sodium ascorbate) to processed meats, such as salami or bacon, to control for potential cancer-causing compounds that might develop once the meat is processed.

In industrial countries, all new food products are assessed for risk prior to being granted approval to be sold in grocery stores. GM foods that are both directly and indirectly consumed are a particular focus of regulatory effort. Directly consumed GM foods include apples, papaya, potatoes, aubergine, sweetcorn, squash, pineapple and salmon. Indirect GM foods would be those that contain ingredients derived from the large-scale commodity production of GM varieties of canola, maize, cotton, sugar beets and soybeans. Canola oil and cottonseed oil are sold directly and also included as ingredients in margarines, salad oils and baking products. Sugars, such as sucrose and fructose, and starches from maize, sugar beets and soybeans are used in a wide variety of processed food products.

All risk assessments are the same and involve scientific assessment of risk thresholds relating to human consumption for allergenicity, toxicity, metabolization, nutrition and dietary exposure. Essentially, this means that new food products are studied to determine how long it takes the product to break down once it enters our stomachs, and whether that may affect the nutrition, function or allergenicity of the food. The more durable the proteins, the more likely our bodies may have some reaction. Our stomach produces hydrochloric acid which is designed to aid in the digestion of our food and to kill any harmful bacteria, such as *E. coli* or *Salmonella* spp., that may be present on our food. Our food safety assessments also look for any potential toxicity. What makes the assessments so challenging is that both dietary exposure and metabolic processes in individuals can vary widely. Consequently, we often focus on subgroups of the population, including those who for economic or cultural reasons consume unusual amounts and those that may be more susceptible either because they are young, old, immunosuppressed or have lifestyles that put them at higher risk. While there is no evidence that any of the new foods developed with new biosciences have had any measurable health effects anywhere, we have a significant number of foods that are part of our traditional diets that expose us to toxins, allergens or anti-nutritional elements. Therefore, food safety and nutrition for new foods cannot be discussed in isolation of existing risks.

Whereas new food products being put on the market are assessed for risks, this is not the case for new processed-food products that include ingredients that have already been approved for safe use and consumption. Thus, the reformulation of an existing food product that switches from one ingredient to another would not be assessed for any aspects of food safety. As long as the base ingredients have safe approval, they can be used in any healthy, or sometimes unhealthy, combination. While most of us prefer to consume healthy foods, we all have a weakness for some special, mostly unhealthy, food product, be it chocolate, ice cream or potato chips. There is nothing wrong in having our own specific ‘go to’ comfort foods, but like all things in life, moderation is key.

## What Is a Natural Food?

The use of the word natural in food marketing is rampant, with a wide variety of definitions and meanings. The origins of the word natural would indicate that the product originated in nature and was sourced from the wild. In our supermarkets today, few if any of the food products we see would look like that in nature. There is a small amount of wild berry picking that occurs, with these fruits being sold in local farmers' markets, often having been processed into jams or jellies by the picker. The sheer volume of demand for the consistent supply of food products has resulted in the domestication of plants and animals, allowing for the consistent production and supply of these products to the grocery stores, either as whole unprocessed products or as ingredients in processed products. So, if products derived directly from nature have vanished from today's global food market, what exactly does natural mean?

Consumers around the world buy into diets that redefine our expectations of food that provide the context for what we think is natural. The oldest and most enduring diets are derived from one's beliefs – Buddhists are vegetarians; Hindus are lacto-vegetarians; and halal and kosher diets avoid pork and prescribe food preparation methods. There are a range of vegetarian diet options, differentiated by what animal or animal by-products one might choose to consume: vegans abstain from any foods produced by animals or tested on animals; fruitarians eat primarily raw fruit; lacto-, ovo- and ovo-lacto-vegetarians accept dairy, eggs or both; pescetarians will eat fish; and pollotarians will eat poultry. Meanwhile, others are less fussy about the source of the nutrition and more concerned about its content, thus we have low-calorie diets (e.g. Weight Watchers), low-carbohydrate diets (e.g. Atkins diet), low-fat diets (e.g. McDougall diet) and a range of crash diets (e.g. grapefruit diet and Beverly Hills diet). Recently, we have added a range of detox diets, which purport to flush out harmful substances by the use of water or juices, and a mix of more quirky suggestions, including that we pair our diet with our blood type or drink less water at meal times. Few if any of these diets are supported by clinical evidence of any differential nutritional benefits, and in a few cases specific harms have been found. Followers of these dietary practices are passionate about their choices, convinced that conventional diets, especially when combined with new or GM foods, contribute to increased rates of cancer, autism, irritable bowel syndrome or Crohn's disease.

At one end of the spectrum, there is the natural food movement, which sells products produced in as natural a state as possible. This segment of the market exploits the appeal-to-nature fallacy, whereby we are led to believe that something that is natural is morally superior or substantively better for us. But as most of us know, and as we will discuss further below, natural things can be dangerous. While providing the impression that the resulting foods are produced without the use of chemicals, this may not be the case, as the natural and organic food industries have a long list of naturally derived chemicals that are allowed in the production of these products. Without the use of these chemicals, yields would either be so low that even the higher prices associated with foods generated through this process would not be sufficient to enable the producer to be profitable, or they would be so damaged by insects that the resulting fruits or vegetables would not be accepted by consumers. Who of us would willingly buy an apple with a worm hole in it, especially if it costs more than prettier, conventional produce? Because many naturally derived chemicals have a lower rate of effectiveness than synthetically developed chemicals, farmers frequently have to make more chemical applications, especially the use of insecticides, to ensure that adequate product is harvested that is acceptable to consumers.

Identifying products as natural has a central role in dairy products. Yogurts are frequently advertised as containing natural probiotics to improve the health of those that consume the product. While these natural enhancements are safe (but arguably not overly effective), there are potentially unsafe uses of natural processes. One unsafe use of natural is in the production of natural or raw milk in milk or dairy products. Natural or raw milk has not been pasteurized and has the potential to contain harmful bacteria such as *Mycobacterium tuberculosis* (which causes tuberculosis (TB)), *E. coli*, and *Salmonella*, *Listeria* or *Campylobacter* species, and a host of other foodborne pathogens. The wholesale pasteurization of milk, combined with better detection and culling of infected herds, beginning in the 1920s, is credited with dramatically reducing the incidence of TB, the single most deadly disease in the history of mankind. In 1900, an estimated 15% of all TB cases in the USA were caused by dairy products. Other pathogens are equally problematic. *Listeria* spp., for example, cause miscarriages, while *E. coli* and *Salmonella* and *Campylobacter* spp. can be fatal to those with weakened immune systems. Through the pasteurization process, these harmful bacteria are killed, providing for the safe consumption of the resulting food products.

In recent years, there has been intense debate within the organic industry as to the use of the word natural, with many companies selling food products that are labelled as natural, but that have not been produced using organic certification standards. Much of the discussion regarding natural relates to the means of field production of crops, focusing on the use of synthetic chemicals and fertilizers. The use of chemicals, such as herbicides to control weeds, insecticides to control insects and fungicides to control plant diseases, contributes to higher yields, thereby keeping food prices lower. Without the use of the crop protection products, yields can be reduced by as much as 60–70%, depending on the crop. Natural fertilizers, such as livestock manure, have been used for millennia as a means of providing plants with the required nutrients. A spate of food recalls and outbreaks of food poisonings involving fresh vegetables in North America and Europe in recent years reminds us that using manure as fertilizer is not without risks. Equally importantly, the availability, quality and supply of natural fertilizers is not adequate for the extensive farming industry that spans the globe. Synthetic fertilizers have been employed for the past few centuries. Chemically, natural or synthetic fertilizers offer the same value to plants and the resulting crops are indistinguishable biochemically.

If the various methods deliver indistinguishable outputs, why is there such an intense marking debate as to what is natural and what is not? The answer is that firms employ the use of natural labelling to differentiate their product, based on unique attributes, from other similar products on the market shelf, all in anticipation of higher sales and prices for the firm. In a multi-billion-dollar industry, a few cents gained on the price of one food product, sold across multiple markets, results in increased revenues in the millions of dollars.

Food companies go to extreme lengths to portray their product as more natural than the similar products of competing firms. Hundreds of millions of dollars per year are spent on the development of marketing campaigns for food products, using farm or livestock images on food products, connections to the names of farms or the endorsement of celebrities. Through the use of product branding, labels are designed to convey to the consumer that a product has connections to nature or natural production, thereby generating some, often mistaken, belief that the product is somehow healthier for them.

Two key elements are vital to the creation of quality in food products: consumer tastes and preferences; and producer capacity to develop consistent, safe, affordable and attractive fare that meets consumer demand. Economic theory suggests that most, if not all, of these elements can and should be produced within minimally regulated markets. Increasingly, variables of trust and confidence in the creation and operation of markets have become the focal point. Markets are often not able to create, by themselves, the conditions of trust that generate the socially optimal quantities of goods and services produced and consumed. Hence, there is a much more explicit role for public and private regulation in markets than neoclassical economic theory generally suggests. This is especially true for GM food products, about which perceived risks and public uncertainties abound.

Although employment of the term natural is meant to convey the impression that the food product in question is safer or better than competing products, there is evidence to the contrary. Conventional and organic crops are far from perfect, and there is no empirical evidence that they are any safer, more nutritious or sustainably produced than GM crops. As discussed elsewhere in this book, GM crops have consistently been found to contribute to less tillage, which increases moisture conservation and reduces soil erosion, while the more efficient and effective use of synthetic fertilizers and less toxic chemicals has delivered higher volumes of quality product. Sometimes these GM products, such as Bt maize, also present with lower toxin levels than conventionally or organically produced crops that are infected through insect activity.

What we can say is that neither process is truly natural. Farming is inherently unnatural. The seeds we have chosen and use, the landscapes we have transformed to our needs and the synthetic or organic chemicals we use are in no way natural. No process can justifiably claim that its products are the bounty of nature, or that it has not had an impact on nature. Nature is only the base for our food system – it is not the boundary for it.

## Science and the Future of Food

The US National Cancer Institute estimates that one in three cancer deaths are diet related and that 8 out of 10 cancers have a nutrition or dietary component. Inverse relationships have been observed between carotenoid-rich foods (which includes carrots, plums, apricots, mangoes, cantaloupes, sweet potatoes, kale and spinach) and certain cancers. Other nutrient-related correlations link dietary fat and fibre to the prevention of colon cancer, folate foods (legumes, asparagus, leafy greens, beets, Brussels sprouts and broccoli) to the prevention of neural tube defects, calcium to the prevention of osteoporosis, psyllium foods (wheat and oats) to the lowering of blood lipid levels and antioxidant nutrients to the scavenging of reactive oxygen species and protection against oxidative damage of cells that may lead to chronic disease, to list just a few. Furthermore, many food components are known to influence the expression of both structural genes and transcription factors in humans.

To improve the nutrition of our staple foods, all breeding techniques need to be applied in pursuit of this goal. Conventional mutagenesis, transgenic techniques and gene editing will all contribute. Conventional plant breeding has been increasing the nutritional content of foods for over 20 years. Table 6.1 provides a summary of some of the nutritional changes that are actively being researched and the related food and livestock crops. Across the seven different broad trait research classifications, 19 different food crops have been involved in research targeted as making nutritional improvements. This provides a sense of scale to the wide variety of food crops that have been involved in targeted research programmes to either improve the nutritional aspects of the product or to reduce attributes that can be harmful to humans.

As plant breeding evolves from random, untargeted mutagenic breeding to targeted, controlled genome editing breeding, there is vast potential to make substantial increases in the nutritional composition of food crops. Not only can these new genome editing technologies increase the healthiness of our food, they will also be able to ensure that some food crops susceptible to disease could be made more resistant. One example is orange groves, which are threatened in many locations by a disease known as citrus

**Table 6.1.** Nutritional traits research.

| Nutritional trait | Crops |
|---|---|
| Increased protein quality | Bahiagrass, canola, maize, potato, rice, soybean, sweet potato, wheat |
| Essential amino acids | Canola, maize, lupin, potato, sorghum, soybean |
| Oils and fatty acids | Canola, maize, cotton, flax, palm, rice, soybean, safflower |
| Carbohydrates | Chicory, maize, potato, rice, soybean, sugar beet |
| Vitamins | Canola, cassava, maize, mustard, potato, rice, strawberry, tomato |
| Secondary metabolites | Apple, alfalfa, maize, kiwi, potato, rice, soybean, tomato, wheat |
| Minerals | Alfalfa, maize, lettuce, rice, soybean, wheat |

greening, which causes the fruit to fall from the tree well in advance of it ripening. Once a grove is infected, it has to be cut down. There is no prevention or treatment for this disease. Similarly, cocoa production is also being devastated by viruses in key production areas while banana bacteria wilt affects the production of bananas and plantains in some key production areas. These plant micro-organisms are most commonly spread by insects and control is currently impracticable. Providing a natural defence mechanism to the plants through improved disease resistance could be an essential part of ensuring that our enjoyment and use of bananas, oranges and chocolate continues far into the future.

The appeal to nature is a trap that may forestall further progress in food security. Because of its intuitive and emotional appeal, it is particularly challenging. History has taught us that the more powerful a framing device, the less capacity we as individuals have to evaluate evidence objectively. In short, passion is a poor basis for planning.

## Further Reading

Block, G., Patterson, B. and Subar, A. (1992) Fruit, vegetables, and cancer prevention: a review of the epidemiological evidence. *Nutrition and Cancer* 18(1), 1–29. DOI: 10.1080/01635589209514201.

McCormick Science Institute (undated) History of Spices. Available at: http://www.mccormickscienceinstitute.com/resources/history-of-spices (accessed 7 February 2019).

Newell-McGloughlin, M. (2014) Health impacts. In: Ludlow, K., Smyth, S.J. and Falck-Zepeda, J. (eds) *Socio-Economic Considerations in Biotechnology Regulations*. Springer Publishers, New York, pp. 125–143.

Niiler, E. (2018) Why gene editing is the next food revolution. *National Geographic*. Available at: https://www.nationalgeographic.com/environment/future-of-food/food-technology-gene-editing/ (accessed 7 February 2019).

# 7 The Local Food Movement: Can We Be Self-Sufficient?

*My family enjoys what has become a Saturday morning ritual – the visit to the local farmers' market. It is a fun experience, enjoyed by us all. Before this outing became a ritual, Saturday mornings were lazy times. They never started early. Pyjamas were often worn almost until high noon. The toaster popped up newly unfrozen waffles (or sometimes pop tarts), the coffee machine gurgled, grapefruit juice and plums were plucked from the fridge, bananas were peeled, and cat food dispensed amid purrs and affectionate rubbing of legs ending in slippers. Junior would be ensconced in front of the kitchen TV watching cartoons so old I thought I remembered them. Alice, now disdaining cartoons as juvenile, would be clutching her phone, texting morning secrets to her gaggle of friends from school and her interpretive dance class. Angela, my spouse, would have her nose in a historical novel – occasionally emerging to google some person or event from the past. Me, I would be in the family room, settling in for a baseball game on the big screen or, if it was not the season for the boys of summer, the highlights from the previous week's sport offerings. These were comfortable mornings but not 'family' mornings – rather it was a time of four solitudes, not that any of us thought something was missing. Change, however, was coming.*

*One October we flew out to spend Canadian Thanksgiving with Angela's sister, Anna, and her family in Ontario. We arrived late Friday evening, but there was no lazy Saturday morning in store for us. Brother-in-law Andy roused us all early, announcing that we were off to the local farmers' market to pick up supplies for the Thanksgiving feast the next day. He stressed that it was important to arrive earlier rather than later, to ensure that the 'good stuff' was still available. It was a three-block walk in the crisp fall air, red leaves for scenery, all seven of us moving together as one family. Elizabeth (no longer Lizzy, as I was forcefully corrected the night before) had been born only a few weeks before Alice. She and Alice were now 'best friends forever'. Our flock rounded a corner and entered the market. It had a certain buzz to it. Anna and Andy were clearly regulars – greetings and introductions were shared, no matter whether it was with those manning the stalls or other customers foraging among them. The coffee stall was the first order of business, as my nose alerted me that the morning jolt of caffeine was close at hand – Andy quickly took orders for a range of beverages, from my black, regular coffee to lattes to mochas to others with long, complicated specifications. Next stop, again directed by the connection between my nose and stomach, the cinnamon bun stall where the warm, gooey, staples (Anna insisted they represent a distinct food group) made for a great breakfast when combined with the coffees – my treat.*

*Elizabeth was busy introducing Alice to a group of (apparently) very cool friends. They moved effortlessly between hanging out together and joining their own family groupings. Junior was actively lobbying for a second cinnamon bun. Anna was consulting her list and strategizing with Angela over which stalls to visit to complete the list. Andy and I were drawn into a lively discussion of the relative merits of Canadian versus American football – until Anna signalled that it was time for the 'hunting and gathering' to begin. Andy and I had central roles as 'beasts of burden' while the children acted as scouts. As we moved through the market Anna and Andy seemed to know all the vendors, who offered friendly greetings and lively discussions of the products displayed. Specials were indicated, samples proffered and appreciated, decisions made and produce packaged up with a flourish. Andy and I were soon laden with the bags we had brought from home, which bulged with an array of the market's best. We had managed to kill the morning. It had been great fun. There were smiles all round as we wended our way back to Anna and Andy's.*

*Thanksgiving over, we returned home. I suggested that we should visit our local farmers' market the next Saturday morning. There was no argument. As our local market was a bit further away, we decided we could take our bikes. Saturday morning arrived, everyone was up and dressed and ready to go in good time. It was a crisp fall morning, blue sky and flat autumn light. The market was humming by the time we parked our bikes. It had the same 'feel' as the one we visited in Ontario. Junior was relieved that there were cinnamon buns. So was I, but I was more relieved that there was an abundance of coffee. Her cinnamon bun and coffee tucked away, Alice soon spied some school friends and was off. Angela, Junior and I began the rounds of the stalls, making acquaintance with the vendors. There were lots of new offerings to sample, produce to inspect, crafts to appreciate and stories to be heard. The farmers were keen to share their tales, the bakers to tout their pies, cookies, cakes and breads, and the artisans to showcase their talent. As we had no bike carriers, we had to walk our bikes home because of all the bulging bags. On the walk home we decided, when possible, to make this a weekly family event. The first order of business was a trip to the bike shop to have carriers fitted.*

*The trip to the farmers' market did become a Saturday morning family 'thing'. We got to know the vendors, the 'regulars' who shopped every weekend, and the secret of the cinnamon buns. I even took a few shifts at a vegetable stall to help during the busy holiday season. Junior's first job was part-time at the vendor that sold cinnamon buns.*

## The Local Food Experience

Over the last 20 years or so, farmers' markets have become very popular – largely for variations on the style of experiences recounted above. There are no definitive numbers on the scale of local food sales, but a cursory survey suggests that traditional markets remain important sources of foods in many developing nations. In developed countries, most of us have come to use the industrial food-supply chain. There has been, however, a push back in recent years, with local food sales rising at least twice as fast as overall sales. As of 2006, the US Department of Agriculture (USDA) estimated that about 3 million Americans purchased food at one of the more than 2400 farmers' markets weekly. Vendors sold about US$1 billion or 0.5% of all farm produce through local markets; although sales by vendors at that point averaged only about US$7100 annually, about one-quarter of those selling through markets derived all of their earnings that way.

There is no doubt that local markets represent a relatively unique experience, combining the satisfaction of a much sought-after pre-modern lifestyle – however idealized – and personal engagement with the local community. There is local produce on offer that is well presented, reflecting the considerable care and attention to detail that went into its production and preparation. The produce is more costly than that sold in supermarkets, which reflects the extra costs associated with that care and attention. The higher cost of food at farmers' markets, however, means that one seldom sees the less well-off members of the community shopping there. Some of the 'regulars' may subconsciously think that is an additional 'benefit' of shopping at the market. The extra cost of food at the market must also cover the cost of running the market because no entrance fee is typically charged to cover the positive benefits provided by the 'market experience'.

One of the other benefits some patrons of farmers' markets perceive is that they are supporting local agricultural production – as indeed they often are. Thus, in their mind they are part of something perceived as a societal benefit – the local food movement. Individuals or families choosing to purchase food from local farmers for whatever reason – lower environmental footprint, supporting the local community, having a higher degree of trust in local farmers – is part of being free to express one's preferences. It is a different matter when fostering local food consumption becomes part of public policy. This is because it may, in turn, lead to drives towards local self-sufficiency, which can put local and global food security at risk.

In most settings, the farmers' market experience only works well if it is buttressed by a well-functioning system of supermarkets and their associated supply chains. Climate and seasonality limit the range of locally produced foods that are available at reasonable prices. For example, neither the coffee nor the cinnamon (in the bun) consumed at the farmers' market could have been 'local' in the vignette that opened this chapter. For that matter, the refined sugar, which is an essential part of a cinnamon bun, is also unlikely to be available locally – even if sugar beets could be grown. The reality is that many 'local' farmers' markets offer a mix of local and non-local products and many products that combine ingredients from a variety of locales. While the carrots, turnips and lettuce may be local, the bananas, dates and blueberries may not be. Local seasons also tend to be short in many areas with these upscale markets. Produce from that local area can only be on offer 'in season'. Out of season, if the particular product is available, it will have its origin elsewhere. If a policy of strict locality of a product's origin were instituted, vendors would have little to offer for much of the year, which would threaten their viability. If the viability of individual vendors is questionable, or they shutter their stalls when local products are not in season, then the viability of the entire market would be at risk. This is one possible explanation for the considerable debate on what constitutes local. Stretching the definition of local can, clearly, increase the potential viability of farmers' markets. A strict requirement that all produce be local may also mean that the viability of such markets becomes much more craft based so that interruptions in production are not a problem, or if for some reason they are a problem, building inventory and storage can be used to bridge shortfalls. In the case of fresh produce, holding and storing inventory is a much more problematic strategy. In any case, while craft-based markets may survive, they will have a reduced place in a consumer or family's acquisition of food.

Supermarkets and their associated supply chains are masters at smoothing over market disruptions, whether they stem from supply disruptions such as crop failures, food safety incidents or interruptions to transport. Their buyers can find alternative sources of supply if a frost disrupts the production of Florida oranges, or drought impacts Californian tomatoes, or grasshoppers devour carrots. It is not often when one goes to the supermarket that what you want is not there. Supermarkets stock upwards of 40,000 different product lines and they are almost all there on the shelves on any given day. This is quite remarkable given that localized and sometimes widespread crop failures are endemic. They happen all the time. While farmers suffer from these failures, consumers in most advanced industrial countries rarely notice them. Supermarket chains work hard at ensuring continuity of supply through diversification of their sources. Local farmers' markets cannot be relied upon to provide the required diversity in their sources of supply. Moreover, it is very difficult to rely solely on a farmers' market to fill one's entire grocery basket. Given how difficult it is to rely on local supplies, even when local is broadly defined, enthusiasm for variations on reliance on local such as the Hundred Mile Diet often wanes relatively quickly.

## The Local Food Movement Exacerbates Food Insecurity

Societies forced to rely on locally produced food are more likely to face food insecurity and market disruptions. No amount of fostering self-sufficiency can ensure food security. Crop failures and other sources of disruptions will happen. Food security is, therefore, at the heart of the debate over the validity of local food as a policy prescription.

For an individual, food security has three important facets: productivity, distribution and access. First, agricultural productivity improvements globally must at least keep pace with total population growth across the world. If this goal is not realized, there will be more and more stomachs that must be fed chasing ever scarcer food. Without productivity growth, food security will be declining for a subset of consumers – likely to be society's poorest members.

Second, once the growth rate in global agricultural productivity exceeds the rate of population increase, food security becomes a question of distribution. An individual's food security depends on food being available, virtually daily, where one lives. When food is not locally available, food security is lacking. If one doubts this, all one needs to do is observe the behaviour of people when a major storm is predicted to hit a local area. Supermarkets are cleaned out by panicked shoppers and sometimes looted. For the most part, these are normally well-behaved people. The disruption to food supply chains in hurricanes, for example, is usually only 3–4 days. Consumers in the local area, however, clearly perceive that they are not food secure.

The final aspect of food security is that even if food is available an individual must be able to access it. Usually access depends on having money to purchase food. It does not matter that the local supermarket, or local farmers' market, is full of food – if one does not have the money to purchase it, that individual is not food secure. Food may be generally available in modern market economies but there are individuals in our societies who do not possess sufficient resources to access it. As a result, there is a place for food banks. Food banks play a redistributive role by providing a mechanism for accessing food that is not based on personal resources. They extend food security to an additional segment of society.

A deterioration in the availability of food arises from three sources. In the first instance, a disruption in local food production can reduce availability. As most of the food consumed in the world is locally produced, crop failures or livestock losses can result in reductions in food availability and increases in the number of individuals that are food insecure. Of course, local crop failures or livestock losses lead to crises in food availability only if the second source of disruption also comes to pass – that food cannot move into areas where there are local shortfalls in production. Local crop failures are manifest all the time in developed countries but go virtually unnoticed by consumers due to the well-functioning supply chains of food retailers that can switch seamlessly to a new source of supply, or that rely on multiple suppliers to ensure constant product delivery. When crops fail, rising prices provide the incentive for food to move from areas where supplies are more abundant. Only if this arbitrage cannot take place, due to limited transportation infrastructure, poorly developed supply chains or barriers to trade, will food availability become a problem. Of course, if food is not available due to a local disruption in production combined with a failure to move food from alternative sources of supply, food security no longer exists. In the worst cases, famines arise. Either people starve or they move to where food is available. One only has to think of the Irish famine of the 1840s, Chinese famines in the 20th century and recent tragedies in Africa – all have meant starvation and populations subsequently fleeing famine areas.

The extreme case of famines deserves further examination. Famines are not static but rather evolve in a dynamic fashion. Famines normally start with a failure in local food production. A rapid rise in the price of food is the immediate effect. If this price increase does not yield food inflows from other locations, food prices will not be competed down to lower levels. Without competition from non-local production, high local prices are likely to be exacerbated by those with the financial resources acquiring and hoarding food – holding food for future personal (or family) consumption – or buying and storing food for speculative purposes – withholding food from the market in the near term with the expectation that prices will be higher in the future. Given high food prices, the poor quickly exhaust their existing resources, acquiring whatever food they can. When their resources are fully depleted, they starve. It is important to remember that in the midst of a famine the poor have no resources to purchase food. Thus, even if non-local food eventually flows into famine areas, the poor cannot purchase it. This was what happened in the Irish Potato Famine – *an Gorta Mór* – in the late 1840s, during which food was latterly available but the Irish peasants had exhausted their savings and had no money to pay for it. It was only much later that the British government was able to restore a reasonable degree of food security when it put in place a mechanism for free food distribution – primarily soup kitchens. Thus, in cases of famine, it is not sufficient simply to ensure supplies of food reach the affected locations; the food must be distributed directly to those lacking the resources to purchase it for the famine to end. The more reliant a community is on local food sources, the less likely it is that the infrastructure for moving large quantities of food from outside the local area will exist. If local food self-sufficiency is promoted through government policy, there is no incentive to develop infrastructure to facilitate inter-area trade. As a result, with the inevitability of a local crop failure, the likelihood of a decline in food security or a famine is increased. Local self-sufficiency cannot be equated with being food secure.

Disruptions to local food production can arise for a number of reasons beyond crop failures resulting from adverse agronomic conditions. Armed conflict can lead to crops being burnt or otherwise destroyed and livestock killed. Further, crops and livestock can often be appropriated by one, or both, of the warring factions, leaving local civilians without access to food. In the wake of such destruction, stocks of seed normally retained from harvests may not exist, leading to a failure to plant. Further, if the conflict leads to lawlessness and, in particular, no capacity to enforce property rights, local farmers may feel it is too risky to sow their crops. Armed conflict can force populations to shift location to escape violence, so that they no longer have access to land upon which to farm.

Armed conflict can also deter those outside the area affected by disruptions to agricultural production from sending their food into a deficit region to take advantage of the existing arbitrage opportunities. Their shipments may be subject to the same potential for destruction or expropriation as local food. Further, even if food could be delivered, local populations may not have any money to purchase it, having already spent it on the meagre supplies of high-priced local food that had been available. It is in these situations that emergency food-aid systems are required. Such systems must have: (i) the funds to acquire food in non-local areas; (ii) the wherewithal to securely deliver food to where it is required and; (iii) a mechanism to distribute the food to individuals in need and who lack resources.

In areas where food is scarce and there are warring parties or a collapse in the rule of law, food-aid shipments, however, can come to have strategic value if they can be stolen from those supplying them. When food is scarce, food stocks can become strategic weapons for those competing for societal control. Once a food-aid shipment can be grabbed by one of the warring factions, it can be used to deny access to those that support the other side or fail to support the group controlling the food with sufficient enthusiasm. Food scarcity, particularly when there has been a failure to plant, can directly affect those doing the fighting. Captured food-aid stocks can be used to feed soldiers and to reward loyalty among supporters. They can also be sold to finance a warring faction's operations. Hence, food-aid stocks become desirable strategic and tactical targets. This means that those supplying food aid, or those that support their efforts, must shoulder the cost of ensuring that food-aid shipments reach those in need.

## Local Food and the Environment

Much is made of the supposed environmental benefit of buying local, based on the lower emissions of greenhouse gases and other pollutants due to shortening the transport distances required to move food to final consumers. The short transport link in the food supply chain is purported to yield a smaller environmental footprint. In some situations, this will be the case. In general, however, this conclusion is based on a naïve view of agricultural supply chains. Full agricultural supply chains do not originate on farm and then extend to the consumer. Agricultural supply chains stretch back to the input suppliers who produce fertilizer, fuel and pesticides. Thus, while the pollution arising from transport may not be fully costed into the private supply chain – there is a negative externality or cost for society – this externality applies through the entire supply chain. For example, if there are no emission controls on truck exhausts, then the firm that runs the trucks pollutes the air making individuals in society worse off without any consequences. Complicating this is that quite often fuel is heavily subsidized in developing countries, which encourages its increased use.

What many may not realize is that it may well create more pollution to move fertilizer, fuel and other inputs to the local farm than the savings that arise from moving the final product a shorter distance. Further, local agronomic conditions may require much greater use of such inputs than were they grown in the highly productive, non-local areas that feed into the supply chains of conventional supermarkets. After all, local products grown using suboptimal scale, technologies and systems are often more expensive than those that are hauled long distances. Even if transport costs do not reflect the full costs to society, the use of other inputs, which often have direct pollution effects, may raise costs due to their relatively inefficient use on farm. Fuel is, after all, needed for tractors and other farm machinery, as well as for transport.

There often are economies of scale, including more efficient energy use in large-scale processing, available nearby, where production can be undertaken efficiently. While some local produce is sold in relatively unprocessed forms, most of the basic foodstuffs that make up the bulk of our nutrition need considerable processing. Local production may simply not be of a scale to justify locating processing plants for each local market. As a result, because locally grown food cannot be processed, this may lead to additional waste as only a small proportion of local production enters the supply chain. The rest may have to be discarded because quality deteriorates without processing. Furthermore, localized harvest periods are often quite short; without the ability to process or export, large quantities of surplus product will drive down seasonal prices to unsustainable levels in local food outlets, creating a seasonal boom–bust cycle that will undercut effective production and markets. Processing allows the bumper supplies during harvest to be levelled out through what is, in effect, storage. Much is made of the food waste generated by supermarkets removing deteriorating, or potentially deteriorating, food from their shelves and disposing of it. However, on-farm waste due to the spikes in supply that come from the immediacy of the need to harvest and no options to process, is often unseen by consumers. Wastage of food – whether on farm, during processing, at the supermarket or in the kitchen – undoubtedly deserves far more attention than it thus far has had; it represents a great waste of resources. We would assert that focus on local supplies would undoubtedly exacerbate the degree of waste.

One indication of the degree of waste associated with local food is simply the quality of the produce on offer. The quality of the produce on offer at farmers' markets draws in consumers – it is so much better than what is sold in the supermarket – and testifies to a great deal of on-farm sorting. Only the best goes to the local market. What happens to the rest? Further, at the end of the day at local markets there may also be considerable waste associated with unsold produce that vendors do not wish to haul back to their farms. It may well not be fit for sale by the time the next market day rolls around. All this needs to be factored into the calculation of the environmental footprint associated with local production.

It is for these reasons that calculating the environmental footprint of agricultural products is very complex – far more complex than simply comparing the transport costs associated with the distance from the farm to the consumer, which is the type of mental calculation made by individuals when presented with the question. This type of back-of-the-hand calculation appeals to our common sense. The reality is that calculating environmental footprints is the realm of experts. Public policy relating to local foods should, hence, be informed by scientific experts.

## Trust and Local Foods

Much is made of consumers being able to trust local farmers, implying that more distant farmers are somehow less trustworthy. I like to be able to look the farmer in the eye. I like to know where my food comes from. Conventional retailers, including both restaurant chains and supermarkets have, in recent years, picked up on this interest of some consumers. As a result, retailers will often feature products prominently in their advertising that can be traced back to individual local farms and local farm families. Of course, it is most often farm families that are identified – rather than solely the farmer. There is an attempt to equate family values with the food on offer. This is clever marketing that plays upon a romanticized, and largely fictional, nostalgic yearning for a past that was less hectic, more wholesome, predictable, stable and generally happier, not to mention substantially more labour intensive for the actual farm families themselves. Farm families are often seen as the personification of that ideal lifestyle. This actually flies in the face of the historical reality – most of the farm population that left agriculture in recent generations did so because incomes were lower and uncertain, farming was very hard work, the risks made life less predictable and for many in North America and Australia the rural life was increasingly lonely. Of course, the lure of the bright lights and superior amenities of city life also played its part.

Now, two or more generations after the great exodus, there is little understanding of the forces that led to the farm population in most developed countries dropping from around 50% of the total national population to 1–5% of the whole. This change is far from complete in developing countries, although, almost universally, developing world cities are swelling due to rapid in-migration from the country, often growing at a pace that defies the ability to provide adequate infrastructure to accommodate the increased population. Few of these urban migrants return to participate in farming again.

Most countries simply let this happen, while some African countries even tax or starve the farmers for capital, accelerating this natural process. China is somewhat of an exception. It has an elaborate system of checks that attempt to slow the pace of urbanization. Whatever the public stance, most migrants find life in the slums and the barrios somehow preferable to the rural poverty and isolation they have left behind. It will be interesting to see if, after a couple of generations, they too will develop a romantic attachment to rural life. These large urban conurbations in developing countries now often outstrip the ability of their hinterlands to supply them with local food – a topic to which we will return.

Of course, every farmer is someone's local farmer. There is no reason to believe that one's local farmer in the Dordogne region in France or upstate New York is any more trustworthy than one in far-away Denmark or California or South Africa or Bangladesh. Reliance on personal relationships as a means to improve the safety and quality of food overestimates the effectiveness of such a strategy – after all, the most effective tactic of all scams is getting the market to trust those perpetrating the scam. It is the oldest trick, from the hawkers of snake oil, the sellers of shares in salted gold mines or shady horse dealers, to modern-day used-car salesmen or actors in infomercials. This does not mean that all farmers are motivated to hide some underhanded practice from consumers, only that some will and being able to look them in the eye offers no protection from those that would. Many a dissatisfied customer was able to look the used-car salesman in the eye yet remained dissatisfied at their resulting purchase of a lemon. The coining of the common phrase *caveat emptor* (buyer beware) suggests that trust in transactions is often misplaced. Local farmers selling in a farmers' market may well be less constrained in their dealings than farmers in formal supply chains where checks on behaviour are built in and commercial penalties can be applied. The farmer whose products are entering a food supply chain halfway around the world is likely to have to meet quality specifications that include on-farm practices that are independently verified; if they fail to supply the contracted quality at the right time and price, they are usually dealt with. Farmers loading up their truck for the trip to the local market face no such constraints. They may face a backlash if caught out by local consumers but the probability of getting caught is likely to be low because on-farm practices are easy to make opaque for consumers and can be explained away as misinterpretations or honest mistakes.

Reports of non-local products being sold as local are sufficiently common to suggest that buyer beware should be routinely applied to products in farmers' markets. The cynicism that is often applied to the practices of large agribusinesses should also be extended to the vendors participating in local markets.

## The Misguided Theory of Self-Sufficiency

An offshoot of the local food movement is the transference of the central idea of self-sufficiency to developing countries. Economists are almost unanimous in discouraging countries, regions, sectors, firms or households from pursuing self-sufficiency as a policy objective. Self-sufficiency in the food space almost always lowers the degree of food security, particularly for those living in the large urban conurbations of developing countries. Many of those cities are more easily supplied through international trade than from their local hinterlands. The infrastructure to move large volumes of local products is often much inferior to that associated with international supply chains.

China, before the opening of the economy to trade in the 1980s, had a formal policy of self-sufficiency. Further, the internal infrastructure was limited in its ability to transfer large quantities of food between regions. This has been widely identified as the reason for the persistence of localized famines. By the early 1980s, Shanghai had reached a population of approximately 20 million inhabitants. The hinterland of Shanghai is relatively productive and does not suffer from major seasonal fluctuations in the ability to produce. Except for the staple rice, storage infrastructure did not really exist. Shanghai's 20 million people had to be provisioned every day from its hinterland. This daily resupply required an extraordinary logistical effort. Each morning, all manner of transport began to wend its way to Shanghai from the surrounding communal farms: horse-drawn wagons, overladen bicycles, trucks of a 1940s vintage, modern trucks, men and women walking with long, bamboo poles slung over their shoulders with baskets at either end, motor cycles, some with sidecars piled high, and slow moving 'iron buffaloes' – small motorized tractors towing wagons of various descriptions. All of these crowded onto two-lane roads moving, for the most part, at the pace of the slowest conveyance. Once the city proper was reached, these vehicles began fanning out into every neighbourhood, every road and every market. At their destination, they were unloaded. If it was watermelon season, trucks laden with watermelon deposited them in large piles on street corners. The effort was enormous. Once unloaded, the vehicles all began their return journey to the agricultural communes. Often, they did not arrive back until the evening. The whole elaborate and energy-sapping process had to be repeated the next morning, and every morning, just to keep the 20 million people in Shanghai fed. It was the epitome of a local food system. There was no parallel, well-functioning system of supermarket supply chains. Local was the only option. It sapped the soul and constrained the productive potential of the city and its population.

Of course, in modern Shanghai all of this is gone. Investments in infrastructure at every level now allow food to be transported efficiently, not only from the hinterland but with imports from outside the region. A deep-water port constructed thirty kilometres out in the Yellow Sea works to facilitate international trade. Storage has lifted the constraint of daily re-provisioning requirements. Shanghai residents have a much greater degree of food security than when they had to rely on local sources of food. Few residents of Shanghai, one suspects, would opt to return to the previous reliance on local food.

As the large cities in developing countries continue to grow and expand – Kolkata, Mexico City, Mumbai, Sao Paulo and a host of others – they will simply outstrip the productive capacity of their adjacent hinterlands to produce food. They must increasingly rely on non-local or imported food. As the global population increases by approximately 35% to more than 9 billion by 2050 – in a little more than 30 years – almost all of those additional people will live in cities, either because they were born there, or they migrated there. They will not be growing food. They will, however, be consuming it. While a considerable proportion of the additional food needed will be grown locally, much will be moved into the cities from the more distant regions of a country or from other countries. Policy makers with their sights on self-sufficiency goals will, counterintuitively, put their peoples' degree of food security in jeopardy. This will also promote the inefficient global use of resources. Imports, by definition, require fewer resources to produce than local production – that is why they are cheaper to acquire.

The desire for food self-sufficiency arises, in part, from a fear of being reliant on non-local sources of food and non-local sellers whom the policy makers do not know and do not trust. Given the diverse sources of supply and the many substitution possibilities among foods, we argue that the potential for more than short-run disruptions to international trade flows are quite remote. This does not mean that such a concern can be totally dismissed. In the food crisis of 2007–2008, several major exporting countries restricted exports to keep prices down for their domestic consumers. This exacerbated the price increases faced by importers and led to declines in food security in a number of already food-insecure, importing countries and, in some cases, reduced the foreign exchange available to finance imports of non-food products that were important to development initiatives.

Instead of attempting to increase self-sufficiency, a wiser strategy is often the strategic use of storage. The food in strategic stockholding can be acquired domestically or from imports. Storage to deal with short-term market disruptions makes good sense. There are two constraints. First, as suggested above, disruptions to local production cannot be avoided. Of course, they also cannot be predicted. Biblical Joseph (of the many coloured coat) might have predicted seven fat years and seven lean years, but there is enough variability that we cannot rely on that rule of thumb. While it may be possible to predict long-term trends, such as the effects of climate change, most local disruptions to agricultural production are the result of short-run shocks – drought, pests, frost or even the violent storms that have become more frequent due to climate change. In the case of the latter we can predict that they will be both more frequent and of increased ferocity, but we cannot predict when and where they will occur. Farmers rightly ask: As I get ready to plant my crop, do I need take into account that my farm will suffer from a violent storm this crop year? It may be possible to predict such weather events 3–4 days into the future but not with enough foresight for actions (e.g. altering planting strategies) to be taken to mitigate the effects.

Second, while increased inflows of non-local products can largely offset the negative effects on consumers of local crop failures, increased inflows of food may simply not happen quickly enough to prevent short-run hardship. Thus, sufficient storage to tide over consumption until additional inflows of non-local food are organized is probably a wise strategy in some circumstances. In developed countries, such cautionary storage is likely to be unnecessary. As discussed above, the supply chains of supermarkets are sufficiently nimble to seamlessly switch to alternative sources of supply and have the logistics capability to move products quickly to address disruption. In developing countries, arbitrage opportunities may not be transparent due to poor communications infrastructure and because transportation networks are weak. In the worst instances of famine, it may be more difficult to find trucks to deliver food aid than it is to acquire the required food. Developing countries are advised by many planners to have 3–4 months of staple crops in storage to allow alternative supply chains to be organized. Stocks can be acquired locally or from non-local sources.

Managing stockholdings is, however, a complex endeavour. An active stockholding programme requires shrewd stock rotation both to minimize physical deterioration of the stocks and to manage financial costs. Stock acquisition must be managed so as not to be too disruptive to consumer markets. Stock releases need to be managed both to prevent stock running out before new non-local supplies arrive and to prevent those in need suffering while stocks are withheld from either the market or as famine relief for those who have already depleted their resources. In the past, these complex tasks have sometimes, maybe often, not been managed well in a number of developing countries. As a result, stockholding has a bad reputation among academics who study food security and, in particular, among policy makers at international institutions. Poor management can be improved, unlike the flaws in a self-sufficiency strategy. Of course, investments in communications and transportation infrastructure may be an optimal strategy, as they would mitigate the need for storage at all.

Instead of having a storage policy that is designed to meet a one-time demand and is a drain on the government budget for stock acquisition, governments may, instead, attempt to increase local production sufficiently to offset any shortfall that may arise from a drought or disease or pest infestation. To do this, they must raise prices for farmers above the level that the market would yield to encourage the additional production. As a result, the subsequent supply will exceed demand at the price paid to farmers and the total crop can be disposed of at whatever price the market will bear. Hence, there is a difference between the price paid to farmers and that paid by consumers – a difference that must be made up by the government on the entire production. This represents an ongoing government expenditure and a permanent market distortion. In the absence of storage, food security is at risk if the degree to which crops fail is underestimated. Estimating the severity of a disruption in production is not an easy task. All in all, storage is a smarter strategy.

The focus on local self-sufficiency by some interested in food policy is difficult to understand. Being self-sufficient makes one vulnerable to natural and manmade disruptions to agricultural production. It is almost as if the idea of famine has faded from memory in the developed world. There was considerable hunger in Europe at the end of the Second World War where local production was largely disrupted as a direct result of the conflict and the strategic destruction of transportation infrastructure. Food supplies had to be shipped from the USA and were gladly received. The local droughts and dust bowls of the 1930s also seem to have been forgotten. Even if memories are short, images of African famines are prominent on television screens across the developed world. One would think we might be more engaged with this vital policy concern as we witness long lines of refugees fleeing areas where local food production has been disrupted or queuing for disbursements of World Food Programme aid in camps.

None of these modern-day events is the result of disruptions to food supplies caused by exporters. They are the result of local food-production failures. Instead of calls for increased trade and the removal of barriers to the transport of food, there are calls for greater self-reliance and self-sufficiency, despite reliance on local supplies clearly having been the cause of food insecurity.

## Winning the Hearts and Minds of Citizens

Of course, the preference for local self-sufficiency has a long history in the protectionist vs open-economy debate that has been going on among economists, and others, since the dawn of economics as a discipline. It was this debate that had a major part in inducing Adam Smith to pen his famous book, *The Wealth of Nations*, in 1776. Of course, *The Wealth of Nations* is considered by many to be the foundation text for economics as a discipline, but the protectionist vs open-economy debate precedes it by a considerable period. The part of Smith's treatise dealing with trade was influential and set the stage for the centuries-long debunking of protectionist arguments. Development and application of increasingly rigorous economic analysis culminated in the general acceptance of the proposition that removing trade barriers is almost always welfare enhancing. No establishment economist would argue against this basic proposition – it is foundational to much of our way of thinking about markets and society. One would have expected that this convergence of professional opinion would have critically, some might say mortally, wounded protectionist arguments. In specialist discussions, that has happened but protectionists still lobby for trade barriers – although for most of us it is clear that they are simply pursuing their own vested interests. Beginning with the repeal of the British Corn Laws in the 1840s and culminating in the WTO Agreement in 1994, we pursued a long, and mostly successful, national and multilateral process of trade liberalization – almost everywhere but for food and agriculture. There, while the arguments for more open trade are generally accepted, removal of trade barriers has proved more difficult, as agricultural protectionists have been much more effective.

This brings us full circle back to the local farmers' market and the local food movement. As we consider where we should direct our societies and economies to move, it is important to separate the positive experience one receives from participating in farmers' markets and local food from what we all really want from our economic and social policy promoted by government. Policy should not be based on the narrow self-interest of wealthy, food-secure consumers in North America and Europe, who, thanks to the agro-industrial system, can afford to romanticize the local farmer. Global food security requires governments to ensure that non-local products can flow between markets without impediments. This does not mean that imports should be given preference or local production should be purposely disadvantaged. If consumers individually wish to express their preferences for local products, and if they are willing or able to pay a premium to express those preferences, who can argue to the contrary.

## Further Reading

Hobbs, J.E., Malla, S., Sogah, E.K. and Yeung, M.T. (2014) *Regulating Health Foods: Policy Challenges and Consumer Conundrums*. Edward Elgar, Cheltenham, UK.

Hobbs, J.E. and Kerr, W.A. (2018) Healthier eating: policy challenges in improving consumers' ability to choose. In: Roberts, D.H. and Josling, T. (eds) *Handbook of International Food and Agricultural Policies, Volume 2: Policies for Food Safety and Quality, Improved Nutrition and Food Security*. World Scientific Publishers, Singapore, pp. 85–109.

Kerr, W.A. (2015) Food security and anti-piracy strategies: the economics of protecting world food program shipments. In: Schmitz, A., Kennedy, P.L. and Schmitz, T. (eds) *Food Security in an Uncertain World – an International Perspective*. Emerald Group Publishing, Bingley, UK, pp. 145–159.

Nakuja, T. and Kerr, W.A. (2018) Food security through public stockholdings and trade distortions. *British Food Journal* 120(5), 1089–1105. DOI: 10.1108/BFJ-07-2017-0391.

# 8 Food Labelling: What Do People Want?

*We don't know about you, but we find labels fascinating. None of us is the primary weekly shopper in our families. When we get sent to the store, we are usually overwhelmed by the choice. We tend to get distracted by the almost limitless choice. The sights and smells of fresh (and sometimes cooking) food excites the senses. Our eyes are drawn to the variety of colour, shapes, sizes, styles and locations of the various products. Even if we go with a list, the eye wanders.*

*The variety is impossible to manage. The average supermarket in the USA now has almost 47,000 items in stock, up from only about 9,000 in 1975. The USDA estimates that more than 20,000 new products are developed and offered in stores annually. It is a moveable feast of opportunity. Selecting what we want and need is not easy.*

*Sometimes, making a choice is easy. I will recognize the size, shape, colour, product name, product picture or some other quick identifier, and grab the item and put it in my cart. If I have time, my eyes will sweep the adjacent shelving to see if there might be something cheaper or more interesting, but for the average person the search and fulfilment for known products can take as little as 20 seconds.*

*If, on the other hand, I am down a foreign aisle, or looking for something not regularly on my list, I am sunk. I get drawn into the labels and product placement notices. Some are like little novels, telling me about a range of things I may or may not care about, usually involving some framing for my choice. At other times, I find labels like a scavenger hunt or puzzle. There are a host of symbols, shapes, colours, words and numbers that are all shorthand for what is in the package, is not in the package or relates to the package in some way.*

*While not everyone reading this will be as naïve as we might be, everyone has had some experience trying to sort out what they need from the choices available. Labels are a key part of that search.*

## Why Label?

The content of food labels is a major flashpoint in the regulation of GMOs. Labels might seem to be a simple fix to the angst among some consumers regarding the use of biotechnology to produce the food they are presented



S.J. Smyth *et al.* (2019). *GM Agriculture and Food Security: Fears and Facts* (S.J. Smyth *et al.*)

with in supermarkets and restaurants. However, the whole question of labelling has widespread implications for everything from international trade to the best use of the space available on packaging – and everything in between. The intent of labels is that they should inform consumers so that they can make better choices – but better means different things to the different actors involved in producing or using labels. For the firm that produces and markets the product, better choices means that consumers will buy more of the product than they would have without the label, and more of the firm's product than that of a competitor. For a government regulator, better may mean that consumers might choose the products that deliver better nutritional value. In this case, there is a clear conflict of objectives and the success of one acts to the detriment of the objectives of the other. Others may propose that labels be used to inform consumers regarding a social issue they may care about – whether child labour was used in production or whether the food was produced sustainably. Again, the objective is to alter consumers' purchasing decisions. There is clearly a lot at stake in the competition over the limited space on labels.

Labels have a relatively unique position in the competition to influence the purchasing decisions of consumers. Consumer and label are face-to-face in the sense that this is the point where the purchaser actually sees the product and the last point that the purchaser can be influenced before purchase. Certainly, advertising and other ways of influencing purchasing behaviour can induce potential buyers to go and seek out a product in a store or to purchase it online, but in many circumstances the consumer may be aware of a product but the decision to purchase it takes place when the label is front and centre. Of course, a consumer can choose to ignore any information on a label beyond the product name; and in some cases, even that information can be ignored if the packaging is distinct (e.g. one knows it is a Coke™ simply by the shape of the bottle even if one cannot see the name). Still, once the potential purchaser has the actual product in front of them, there is a chance that label information can influence whether or not the product is purchased.

The primary purpose of labels is to identify what economists call credence attributes of a product. Products can be divided into three types based on their attributes: (i) inspection or search products; (ii) experience products; and (iii) credence products. Inspection products are those whose attributes can, for the most part, be determined at point of purchase. The quality of fruit can be assessed in the market at point of purchase. Most fresh fruit and vegetables are purchased in this way after potential purchasers inspect them at the fruit stall in a market or the supermarket. These products are not generally labelled individually, but at the bin level.

Experience goods are those whose quality cannot be determined through inspection but only through the experience of consumption. While one can inspect a steak in a supermarket, it is not possible to determine the tenderness of it until one consumes it. Through experience, one may be able to learn that this particular cut of steak is more likely to be tender.

With credence goods, a product's attributes cannot be determined either by inspection or experience. Even after drinking a coffee, it is not possible to know whether it was produced using child labour. Even after inspecting a mango at a fruit stand and eating it, one cannot know whether it was produced organically. Having both inspected and consumed a potato, it is not possible to know whether it is GM or not. If consumers are to be informed that a product possesses a credence attribute, it must be labelled. There is both a physical and psychological limit to labels – there is no room for an infinite array of labelled attributes given physical constraints, while meaning is lost if too much is communicated. Therefore, the amount of information must be rationed by some method. This raises questions regarding what is important and who makes the decision?

## What Should We Label?

Food products are biological in origin. This means that they are not uniform in their quality; there will always be biological variation in food products. The human consumers of food are also biological entities and exhibit an astonishing degree of biological diversity – as is said, we come in all shapes and sizes. The efficacy and functioning of our biological systems also vary considerably. Some individuals have difficulty managing blood-sugar levels, some have peanut allergies, some poorly process gluten and some have a predilection for alcohol dependency. The needs of a human body also vary over a lifetime. The food requirements of a teenager are dramatically different from someone in their 80s.

The result is that food consumption is the matching of a large number of foods that exhibit an almost infinite range of quality with more than 7.5 billion unique biological individuals. Add in the predilection of humans to desire different lifestyles – to run marathons, to be skinnier, to be vegan, to avoid spinach, to consume chocolate on a regular basis, to fast, or to eat hotdogs or shelled peanuts at baseball games – and dietary needs and preferences will often require unique matching solutions. Layer on top of those nutritional and lifestyle facets of food consumption social choices, such as unwillingness to consume foods produced using prison labour, derived from animals slaughtered in particular ways, or grown in countries with a particular political system, and one has a large need for information pertaining to food. Given the credence attributes of foods and the interests of consumers, labels should almost be unique to an individual. The label information of interest to an 80-year-old, Jewish, wheelchair-bound woman is entirely different from a 20-year-old woman in the Arizona desert training for a triathlon. Each needs a unique set of information on available foods if they are to maximize the value they can obtain from their food consumption.

Ask yourself what information you personally would like to see on a food product – don't think about the information that is currently on labels but rather what you would like to see. Do you think this would be the same for all members of your family? How does your list match up with the information currently found on labels? Is there information currently on labels that you are not interested in or don't think is useful? Beyond identifying the product, do you read the labels at all? Do you read the label carefully and digest all the information the label provides, or do you simply look for certain key information and/or symbols? If the information you would like is not provided, does it deter you from purchasing the product? Do you actually understand what the label says?

Of course, individually tailored labels are not feasible because the provision of information is not costless and the ability of individuals to process information is limited – economists have found most of us are 'boundedly rational', in that we don't want to know everything about what we buy (or don't buy). Sometimes more information simply makes us less happy. Given that there is, however, a very large interest in credence attributes and a limited space to provide those on a label, information on which attributes should be included on a label becomes an important question. As suggested above, who makes those choices is also very important.

## Who Gets to Set the Label?

There are normally two parties involved in putting information on labels. First there is the firm that produces or processes a food product in a later stage of the product's supply chain – most often the firm that packages the product prior to retail. Second, there is the government that regulates the provision of public-interest information on labels. The government may either require information to be included on labels or refuse to allow some information to be included on labels. Other actors, such as NGOs, may lobby either the private firm or the government for inclusion of their information on a label. For example, a group that provides halal certification may approach a meat processor to include their symbol on packages of products that they have certified. This may be mutually beneficial, in that customers desiring halal produce will have reduced search costs and the firm selling the meat product could experience increased sales. Public health NGOs may lobby governments to require certain nutritional information to be included on food products.

A distinction is usually made between voluntary and mandatory labelling. Firms make voluntary decisions regarding the information they wish to include on their labels. These decisions are motivated by commercial considerations – will the information increase revenues more than the cost of providing it? Additionally, will the information offered add more to net revenue than alternative information that competes for the limited space on the label? Governments use mandatory regulations to ensure that appropriate information is included on labels, such as ingredients, nutritional values, weights or measures and the name of the firm producing the product. The aim should be to enhance the public good. A typical food label carries a combination of voluntary private information and information mandated by the government. Whether by negotiation between the representatives of private industry or some other means, an accommodation has to be reached regarding the proportion of private and mandated information on labels. Given space limitations, both governments and the private sector face choices as to what information appears on labels.

Voluntary private decisions regarding the information provided are based, as suggested above, on commercial considerations. These commercial decisions are further guided by the effects of aesthetics such as visual appeal, crowding and the potential for information overload or fatigue. Crowding can tax consumers' patience by forcing them to search for the particular piece of information that they have an interest in. A label with multiple symbols, logos and printed information may not serve the uptake of the information that its inclusion on the label hopes to impart. Label designers must also accommodate what the government has mandated for inclusion. Regulators may also mandate the form the information must take and how it must look. Label designers must work around the constraints imposed by the mandates to arrive at a scheme for the label that will still appeal to potential customers. Governments may also mandate approvals for labels before they can be applied to products. The approval process can be a source of delay and constrain the ability of food companies to capitalize on short-run consumer trends pertaining to the information they wish displayed on labels.

Mandating information is a complex process. The objective is to ensure that the information provided on labels allows consumers to make better choices. Many countries mandate that nutritional information be included on labels – but what information will be the most useful? In what format should the information be provided? Text and numerical information may be too complicated for some consumers to understand. Even if the information is provided, many consumers may not have sufficient understanding of nutrition to benefit from it. The cost of providing the nutritional information mandated may be expensive for the firm, particularly for products with multiple ingredients. Labelling may also restrict the flexibility of firms to respond to markets. A firm making granola bars, for example, may wish to change the mix of grains and other ingredients as the relative prices of these inputs fluctuate daily. Their labels may, for example, require ingredients to be listed in order of importance in the final product. Further, the nutritional content of the final product may change with the least-cost mix of ingredients. This would involve an expensive retesting for nutritional content, a change in the nutritional information on the package and approval of the altered label. Faced with this cost and delay, the firm may choose not to alter the composition of their granola bars. In the process, however, consumers will be denied a less expensive product.

The process of devising nutritional labels is not a simple or straightforward one. If it were, one might see a convergence in labelling requirements internationally. This has not been the experience. Different countries require different information in different formats on their labels. These differences can affect international trade. If a potential exporter's required domestic information differs from that of the nutritional information required by the importing country's regulations, then imports may not be allowed. In such cases exporters will have to redesign their labels, retest their products to provide the information required by the importer and reconfigure their production lines to incorporate the alternative labels. Approvals may have to be sought from the foreign regulatory authority for the exporter's proposed label – delays in label approvals can be an effective barrier to trade. Differences in labelling requirements will complicate inventory management as separation of inventory will be required. The more foreign markets a firm wishes to serve, the greater the costs will be.

Given the costs associated with divergent nutritional labelling requirements among countries, it is somewhat surprising that there has not been more international standardization of nutrition labels. That differences persist suggests just how difficult it is to find agreement on the content and format of labels – both domestically and internationally.

Beyond attempting to provide information on credence attributes of individual foods, government regulators must also protect consumers from hazards and fraud. Hazards arise from credence attributes that can be harmful for some. Credence attributes that can be detrimental to the health or wellbeing of all, or nearly all, members of society lead to products whose sale is banned – hence there is no need for labelling. Hazard labelling arises when the credence attribute is detrimental to a subset of the population. Banning peanuts from sale would be too restrictive on consumers because only a segment of the population is afflicted with an allergy to peanuts. Peanuts are only a hazard for some, therefore the presence of peanuts should be mandatorily labelled to protect those who have a peanut allergy.

In cases of hazard for some, regulators are still faced with the constraints associated with space on the labels. Allergies arise for some individuals over a wide range of credence attributes. Given the space constraint, not all potential allergens can be labelled. Thus, unrecognizable hazards will continue to exist for some consumers. For example, many markets allow processors to group all spices in a generic declaration, even though some are allergic to specific spices, such as the ubiquitous mustard. Regulators face difficult decisions regarding the cut-off point for which hazard information is included on labels and which is not. Regulators are likely to choose those to be labelled based on the prevalence of the allergy in the population, the vulnerability of subsets of the population having the allergy (e.g. children), the severity of a typical allergic reaction and/or the likelihood of the allergen being present in the food. In any case, one will feel ill-served if one's particular allergy is below the cut-off for mandatory labelling, as you will be vulnerable in a way that someone with an applicable hazard label is not.

Protecting consumers from false or fraudulent information is another role for mandatory labels. Horsemeat should not be allowed to be labelled as beef, or lettuce labelled as being organic when it was produced conventionally using synthetic pesticides and herbicides. One of the more contentious areas of food labelling is in the area of health foods or nutraceuticals. These are foods that are marketed as being superior to unenhanced varieties (e.g. omega-3-enriched eggs compared with normal eggs). What claims should those marketing enhanced products be allowed to make? If no regulation takes place, fraudulent claims can be made to enhance sales. 'Buyer beware' applies. While this can work as a regulatory approach, in many areas of the economy governments choose to proactively protect consumers from fraud. In the case of health foods, it may be difficult for individual consumers to discern the efficacy of the enhancement. How can an individual know whether the claim that the enhanced product reduces the probability of liver cancer occurring is legitimate? With no information regarding efficacy, a consumer may be wary and not purchase the product though it actually has a high degree of effectiveness. The result is suboptimal from society's perspective. Alternatively, the claim may be false, and the consumer wastes their resources purchasing it. Those resources might have been spent more effectively on alternatives that would actually reduce the probability of contracting liver cancer.

If regulators decide that it is in the public interest to regulate the claims made on products, then the question of proof of efficacy arises. One route would be to apply the standards for proof of efficacy required for new pharmaceuticals. To be able to have a new drug approved for sale and its function advertised requires very extensive, rigorous and costly testing to be undertaken by the pharmaceutical firm, including randomized controlled trials using human subjects. The health-food industry has, for the most part, resisted this very demanding level of proof, primarily due to its very high cost. The health-food industry argues that the high cost of proving efficacy will deter firms from developing products and, hence, deny consumers access to products they would have an interest in. The growth of the health-food and associated industries has been rapid, even without any compelling proof of efficacy. This suggest that consumers may not be as concerned with proof of efficacy as they may be with safety. Resistance to providing proof of efficacy is seen by many simply as an opportunity to hoodwink naïve or sometimes desperate (for a cure) consumers – making profits out of credence attributes that are claimed but do not actually exist (or cannot be proven to exist). References to the snake oil salesmen of the unregulated frontier era in the USA are often made.

Without rigorous scientific evaluation, determining the appropriate evidence to make claims on labels has proven to be a challenge for regulators. Should a health-food firm be allowed to claim that its product reduces heart attacks, or reduces the probability of heart attacks, or is good for heart health, or is endorsed by cardiologists, or taken for heart health by an Olympic athlete, or ...? Are scientific studies that indicate an association between reduced heart attacks and consumption of a health food with a particular credence attribute, but without clinical trials, sufficient evidence to allow claims on labels? Are testimonials by doctors and other healthcare workers sufficient evidence? Are testimonials by people who consume the health food and subsequently feel better sufficient to allow a claim (e.g. over 100,000 satisfied customers). Regulators in different countries have, unsurprisingly, come to different conclusions and approaches to allowing claims on labels. As a result, internationally the same product may have very different claims on its labels in different markets.

At the most basic, less is more. Too many words can crowd a label, which may discourage consumers from reading it. The often-technical jargon on a label may not impart any information at all to the user. To overcome these problems, symbols rather than words are preferred for labels – in effect, we use heuristics to signal quality attributes. Symbols may be more easily identified at a distance as shoppers trundle around the aisles of supermarkets. A bright red heart or a distinctive green logo can easily catch the eye of the shopper where words cannot. The symbol is only effective if the marketer informs consumers about the symbol and its meaning. However, the regulators are very concerned about the message that is conveyed to the consumer. Does the heart on the package convey: the product reduces heart attacks, or reduces the probability of heart attacks, or is good for heart health, or is endorsed by cardiologists, or taken for heart health by an Olympic athlete, or …? Heart health has sound science behind it. It is far more complicated to convey sustainability. What does a logo associated with sustainability actually mean and, more importantly, how is it interpreted by consumers?

## Monitoring and Enforcing Labels

Clearly the questions pertaining to government-mandated labelling regulations are complex. They go far beyond decisions solely related to what appears on a label. As the products that require mandatory labelling often involve credence attributes, there must be evidence that they actually occur in the food product. For example, if a government decides that it wishes to regulate claims pertaining to gluten for public-health reasons, and to allow a gluten-free claim, then there must be monitoring mechanisms put in place to ensure that the claim is accurate. For those who pursue a gluten-free diet as a lifestyle choice, this may be enough, but for those with coeliac disease, who would be harmed by inadvertently ingesting gluten, there must be mechanisms put in place to have the product withdrawn from the market if the claim is found to be untrue and gluten is present in the product. Individuals with coeliac disease need to have confidence in a label claiming the product is gluten free.

Assuming that a product that can claim to be gluten free can command a higher price than a similar product that contains gluten, unscrupulous food producers would be tempted to simply put a gluten-free label on their product that actually contains gluten. They would then receive the higher price without providing the credence attribute. In some cases, determining the source of the gluten which caused a reaction in coeliac sufferers is straightforward, but in many cases it is not. Individuals draw their diet from many sources and there are often lags before a reaction is observed. Before the state could intervene in the market, it would need scientific proof that the suspect products were improperly labelled. They would first have to be identified and then tested for the presence of gluten – all of which takes resources and time. All the while, sales would continue, the price premium would be received and those with coeliac disease would suffer.

One has only to point to the difficulty Chinese authorities had in identifying and removing from the market baby formula and other milk-based products that had been adulterated with melamine by a number of Chinese dairy-product firms. A range of milk-based products were purposely adulterated with melamine – a type of plastic that can falsely indicate higher protein levels in certain foods but that can cause renal failure and kidney stones when eaten by humans. Before the source was determined and dealt with, it has been estimated that there were more than 300,000 victims. Over 50,000 babies were hospitalized. Six infants died.

The combination of unscrupulous business people and honest mistakes suggest that monitoring products where claims on labels are allowed is only prudent. The supply chains of food products are often long and complex, with products passing through many hands in their progress from the farm to consumers. Protocols and checks must be put in place to ensure that, for example, a product containing gluten is not allowed to enter the supply chain by accident or design. To prevent purposeful adulteration, testing along the supply chain must take place. There must be an ability to segregate gluten-free ingredients and products from those that contain gluten. One complication is that many production, processing, storage and transportation facilities are not designed for a high degree of segregation. Hence, ensuring that contamination of product claimed to be gluten free by products containing gluten can only be done at a reasonable cost to a non-zero standard. In the case of gluten's threat to coeliac sufferers, the degree of segregation that is commercially attainable is sufficient to remove the threat of adverse reactions. Zero contamination is neither needed nor achieved.

The case of peanuts illustrates the difficulties associated with effective segregation and labelling. Peanut allergies are widespread and can be deadly for some. As a clear case of a hazard for some, many governments have decided to require labelling. There may be a direct warning on the label regarding the hazard or ingredients must be listed. Allergens are sometimes emphasized in the list of ingredients. As the threshold for consumption without an allergic reaction is very low, accidental contamination along supply chains is a real problem. As a result, products that do not include peanuts by design may sometimes inadvertently contain peanut fragments. Food companies can go to great, and costly, lengths to ensure segregation, including segregating entire production facilities in attempts to ensure no contamination. Due to the catastrophic effect of ingesting peanuts for some, failure to label the presence of peanuts when they exist in the food product can have significant legal ramifications. In effect, not listing peanuts as an ingredient on a label is tantamount to claiming that the product is peanut free and safe for those with a peanut allergy to eat. As a result, it is commercially prudent to label the presence of peanuts even after having done everything commercially possible to ensure segregation – because ensuring the zero presence of peanuts is not an attainable goal.

The labelling solution in some jurisdictions is a 'may contain' claim, used to warn those with peanut allergies of the risk. It is legally prudent to put the 'may contain' label on a wide range of products that may well have a very low probability of contamination and, hence, be safe to eat. That can, however, effectively narrow food choices for some consumers to a radical degree.

For some with peanut allergies, avoiding all products which carry the 'may contain' label is an unacceptable restriction on their food choices, and they sometimes choose to ignore the warnings on the label and take the risk of a reaction. In those cases, the label is not useful, in that excessive, and in some cases improper, use of the 'may contain' warning reduces its effectiveness.

The costs that can be imposed on the food industry from labelling requirements can be illustrated by the North American experience with country of origin labelling (COOL). Some consumers may well have an interest in the national origin of the products they are consuming. They may have a preference to consume products originating in their own domestic market. COOL is justified on the grounds of consumers' right to know as well as being a measure to prevent fraud. If such preferences are sufficiently strong then domestic producers have an incentive to voluntarily label their products as being domestically sourced – 'Product of the USA', for example. Voluntary labelling, however, does not prevent foreign producers from mislabelling their products 'Product of the USA' as well. Voluntary labelling also does not deal with those foreign firms that simply do not identify the national origin of their products, thus depriving interested consumers of their right to know a product's origin.

The right of countries to insist on COOL was agreed in the General Agreement on Tariffs and Trade (GATT), negotiated in 1947, and continues in the WTO of 1994, albeit with certain conditions attached. The US Congress in 2002 passed specific COOL legislation – sometimes referred to as mCOOL, with the m standing for mandatory – which came into force in 2008. While the law applied to a range of fresh fruits, nuts and vegetables, its primary target was fresh beef and pork. In the wake of the North American Free Trade Agreement (NAFTA) of 1994, the North American livestock industries became highly integrated. Part of that integrated trade flow consisted of cattle or pigs born in Canada and shipped live to the USA for fattening or slaughter. Those promoting the legislation in the USA hoped that US consumers would respond positively to having information on the origin of fresh beef and pork and buy more US product.

The problem from the perspective of the US beef and pork industries was the integrated structure of their industry, involving producers in Canada, Mexico and the USA. The legislation specified that to be identified as US origin the animal that was the source of the meat had to be born, raised and slaughtered in the USA. If the animal was born and raised in Canada and sent to the USA for slaughter, the label had to reflect that fact. Further, if an animal was born in Mexico but was raised and slaughtered in the USA, the label had to reflect that. The effect of these labelling requirements was that animals had to be segregated in production and processing as well as during transportation in order to ensure that the label on meat cuts was correct (so that no meat of mixed origin would be mislabelled as originating in the USA). For many firms, the costs associated with segregation were prohibitive and they decided to no longer accept cattle or pigs from Canada or Mexico. Meanwhile, to offset the costs of segregation, Canadian and Mexican animals received heavily discounted purchase prices from US buyers. Hence, the labelling requirement acted as a barrier to trade. Some speculate that those proposing and supporting COOL hoped the labelling would provide a measure of economic protection from imports.

In the end, the trade-restricting aspects of the COOL regulations led to it being found in contravention of WTO commitments and eventually being rescinded by the US Congress. It is clear that changes in labelling can have significant economic effects and hence legislators and regulators need to understand the ramifications of requirements they put in place.

Voluntary labels can also result in considerable costs in terms of verification and monitoring. A firm may wish to voluntarily label their products as being produced in an animal-friendly fashion, or as being produced sustainably, or that farmers receive fair prices or farm workers receive fair wages – usually called fair-trade products. If one is going to make such a claim voluntarily on a product label, then one should be prepared to prove that is the case. There are many incidents in which firms have been accused of green washing, whereby they claim to be behaving in an environmentally friendly fashion when in reality they have not fundamentally changed their practices. Other firms have claimed and labelled that their products are not produced using child labour, prison labour or slave labour, only to have their supply chains inspected and those labour practices found to exist among their suppliers. Being found not to be living up to one's label claims can be very damaging to a firm's reputation. It may take considerable resources to ensure that one's claims on voluntary labels are being fulfilled. Claiming that workers are well treated, for example, under fair-trade commitments may require standards being established for all manner of practices related to housing, the schooling of children, healthcare, the provision of safety equipment, sanitation, and maternity leave, among others. All of these must be developed and monitored to ensure that they are strictly followed by managers, franchise holders or supplying firms. In some cases, the specific standards that must be complied with, and hence monitored, run to many pages of detailed requirements with accompanying boxes to tick. All of this to ensure that what the label claims has actually been undertaken. If a firm is caught out not complying with its own claim, considerable damage to its reputation can follow.

## Labelling for GMO Foods

As suggested at the beginning of this chapter, one of the most contentious issues surrounding the labelling of food products in recent years has been whether or not the use of GMOs in agricultural production should be reported on the resulting food products – either as a mandatory or voluntary measure. The issue of GMO labelling is part of a larger issue over the need for scientific justifications for denying technologies. The issue of how to regulate GM technology has played out most prominently in the international trade forums, where conflicting regulatory approaches result in trade barriers. The major protagonists on each side have been the USA, which supports what has been described as a scientific rationality approach, and the EU, which pursues a social rationality approach.

The scientific rationality approach followed by the USA and a number of other countries follows a logic that accepts that progress is a function of technological improvement and, hence, a technology should be rejected only if there is a scientific reason that it should not go forward. In other words, the technology would need to represent an unacceptable risk to the society. This then drives the formal risk analysis framework that has been developed to assess the acceptability of new technology solely from a scientific perspective. While the underling risk analysis framework is used extensively in countries that follow both a scientific rationality path and a social rationality path, in the case of countries following the scientific rationality approach this process makes the final judgement on the acceptability of the technology. This is not the case when a country espouses social rationality. In those cases, the scientific assessment only informs the decision process pertaining to technology. If the scientific assessment is negative, then just as in the scientific rationality approach, the technology should be rejected. Passing the scientific rationality assessment, however, does not mean the technology would necessarily move forward in countries espousing social rationality.

Under the social rationality approach, new technology is seen as a disruptive force that disturbs the socially accepted norms and thus needs a broader buy-in by society. Thus, once the scientific acceptability hurdle has been passed, then other concerns need to be taken into consideration. How this is accomplished is usually not transparent, as it involves politicians and other interested parties making a range of subjective trade-offs.

As new technologies always create both winners and losers, politicians are pushed and pulled to manage the impact. For that reason, some nations like the scientific rationality approach as it effectively de-politicizes technical progress. In most cases, scientifically rational process will not deny market access to a safe and efficacious technology solely because it creates losers. These governments may still wish to prevent losses from arising, or to compensate losers, but they do that through the political process only after adoption is approved.

In the case of agricultural biotechnology in both scientific rationality countries and social rationality countries, the scientific safety and efficacy hurdles have often been cleared easily. In countries following the social rationality approach, however, the acceptance of GM technology has stalled. In the EU and a number of other countries, GMO approvals are very slow, expensive and unpredictable, if they even occur at all. Further, those systems seem to be engaged in perpetual policy reform. In the EU, for example, after more than 15 years of review and debate, the EU reformed its system and restarted EU-level approvals for new GM crops, but individual member states asserted their right to ignore the approvals. In effect, new GMOs have been banned for cultivation and human consumption since the end of the last millennium for social and not scientific reasons. There are small areas in the EU where GM crops are grown, currently only Spain and Portugal, but the approval of these GM crops pre-dates GMOs having become a major political issue and the resulting ban on new approvals. Despite having almost no GMO products on the market, the EU has a strict policy of mandatory labelling, much of which is moot because many GM ingredients are simply not available in the community.

The fight over GMO labelling has largely taken place in countries that follow the scientific rationality approach. There are 28 countries where, following GMO cultivation approval (1995–2017), GMOs are widely available in the market. Anti-GMO activists and those who have concerns regarding consuming GMOs have targeted those markets, advocating mandatory labelling for GMOs in foods. They justify their advocacy on the basis of a consumer's right to know. Supporters of the scientific rationality approach have argued against mandatory labelling of GMOs, arguing that there is no evidence of potential harm from consuming foods that are derived from GMOs.

The argument against mandatory labelling of GMOs is that governments should not be adding information when there is no scientific reason for adding it – there is no hazard to justify a label. They argue that mandatory government labelling should be reserved for actual hazards – in effect reserving the warnings for when they are most needed. They believe that this is what consumers expect from regulators.

If consumers truly only expect governments to label real and imminent hazards, then requiring the labelling of GMOs would naturally cause some consumers to ask the question: What is wrong with this product if the government has gone to the trouble of requiring GMOs to be labelled? Is there some hazard associated with this product? The government would not label the product if there is not a problem with GMOs, would it? If consumers react in this way to a mandatory GMO label, they may be less likely to purchase the product. This puts those firms wishing to sell foods produced using GM technology at a commercial disadvantage relative to those selling conventional (non-GMO) products. They argue that there is no public-interest justification for them to suffer from mandatory labelling and no reason to raise consumer angst. They note that those consumers with a strong preference for non-GMO food can easily find what they want in the organic sector. If demand exceeds the capacity of that sector to supply it, then there is nothing to prevent some entrepreneur developing a GMO-free supply chain and voluntarily labelling the new supplies as GMO free.

Those advocating for mandatory labelling do so on the grounds that they have a right to know whether or not GMOs were used in the production of the food they consume. As the use of GM technology is a credence attribute, they have no way of knowing whether the food they are about to purchase is GMO or not. They may wish to have the choice to not purchase food that used GM technology because they reject the scientific approval – the methods or thresholds for approval were not acceptable to them. Alternatively, they may accept the scientific approval but still wish not to purchase food that used GM seeds in production because they have a moral or religious reason for rejecting the technology. Other groups may be concerned about the large agribusinesses that have developed and commercialized GM crops having too much influence in the food system.

This has come to the fore in discussions pertaining to processed agricultural products such as the oil derived from GM canola. The processing of canola into oil eliminates any trace of GM material from the oil, such that there is not even a potential unrecognized hazard. Many argue that this negates any need to label the product. Others argue that labelling should be required because their objections stem from grounds other than a potential unrecognized hazard. They cite moral objections to unnatural gene transfers or the role that large agribusiness firms play in the sector.

It can be argued, however, that not labelling GM crops and foods can also have a negative commercial impact. If a significant proportion of consumers have concerns about GMOs – sufficient concerns that they wish to avoid GMOs in their food – then they may avoid purchasing an unlabelled food product altogether. If they know that GM maize exists and is in foods being sold but cannot tell if the food with maize as an ingredient is GM or not because GMOs are not labelled, they may choose not to consume maize but rather to consume wheat, for which they know no GM crop exists. This lack of transparency will negatively impact GM and non-GM maize producers equally.

Those who advocate for mandatory GMO labelling are likely to expect that once labelled, consumers will shift to products not labelled as GMO. They expect that consumers will be willing to pay a premium for these products and that the price of GMO-labelled products will fall. This result, however, is far from assured. While producers using GM technology will resist the imposition of mandatory GMO labelling, once labelling is mandated, they have a relatively simple task.

Putting labels on GMO products is easy – just slap a label on the final product in accordance with the regulations. Even if the labelling process is imperfect and some non-GMO product is mislabelled as GMO, no one will care. Those marketing GMO foods incur no segregation costs. Choosing not to put the mandated GMO label on a product de facto means that one is claiming that one's food product is GMO free. Hence, if a firm wishes to market its product as being non-GMO then it will have to take measures to ensure that the non-GMO status is maintained throughout the entire length of the supply chain. This is because consumers interested in purchasing non-GMO foods will care if food labelled non-GMO has GMOs mingled in it. As discussed above, segregating supply chains is a major challenge for food companies. All other things being equal, the costs associated with segregating non-GMO foods will cause the price of non-GMO products to increase. In contrast, those products having to be labelled GMO will not have an increase in costs, so prices should remain stable. As the price of non-GMO foods climbs relative to GMO foods, the quantity of non-GMO food demanded is likely to fall. Some consumers who would like to consume non-GMO foods will find them too expensive and switch, maybe reluctantly, to GMO foods.

If the costs of segregation prove to be prohibitive, then one might see a replay of the peanut labelling regime. Those marketing non-GMO food may resort to labels saying may contain GMOs. In the end, the result of mandating the labelling of foods where GMOs were used in production may be labels that do not actually provide useful information to those that wish to avoid consuming GMO food products. This is hardly the result desired by those pushing to have mandatory GMO labelling.

If there is no hazard associated with the use of GMO technology, there is a legitimate question of whether mandatory GMO information should have priority over other information competing for the limited space on labels. For example, allergy alerts for peanuts are mandated in many jurisdictions because of the hazard peanuts pose for some consumers. Other consumers have allergies arising from cashews or red seafood, which are not currently mandated on labels, yet pose real hazards to some consumers. Should non-hazardous GMOs receive space on labels in preference over the real hazards associated with cashews? If labelling requirements are to be expanded, what should be the criteria for ranking which potential labels receive space? In some jurisdictions, GMO labelling has been the focus of ballot initiatives that are voted upon. A consumer voting to have mandatory GMO labelling seldom has the entire picture about labels and their impacts. Given full information about the space constraints and alternative credence attributes that could also be labelled, many might think differently. Voting initiatives may be popular with single-issue advocates but they are not a reasonable way to determine labelling policy, if our goal is to assist consumers in accessing safe and nutritious food.

## Labels for What and for Who?

Labelling is often put forward as a simple solution to the problems associated with credence attributes in food. While labels may be central to addressing the credence nature of many attributes of food, their use is seldom simple in practice. Regulators face difficult choices in what to label and implementation often takes considerable resources. As the economic effects of labels work their way through their impact on consumers and supply chains, the results often are not as anticipated. Hence, labelling should be approached carefully by firms, policy makers and others in civil society with an interest in assessing the quality of the food consumed.

## Further Reading

Gaisford, J.D., Hobbs, J.E., Kerr, W.A., Perdikis, N. and Plunkett, M.D. (2001) *The Economics of Biotechnology*. Edward Elgar, Cheltenham, UK.

Hobbs, J.E. and Kerr, W.A. (2006) Consumer information, labelling and international trade in agri-food products. *Food Policy* 31(1), 78–89. DOI: 10.1016/j.foodpol.2005.08.001.

Hobbs, J.E., Malla, S., Sogah, E.K. and Yeung, M.T. (2014) *Regulating Health Foods: Policy Challenges and Consumer Conundrums*. Edward Elgar, Cheltenham, UK.

Hobbs, J.E. and Kerr, W.A. (2018) Healthier eating: policy challenges in improving consumers' ability to choose. In: Roberts, D.H. and Josling, T. (eds) *Handbook of International Food and Agricultural Policies, Volume 2: Policies for Food Safety and Quality, Improved Nutrition and Food Security*. World Scientific Publishers, Singapore, pp. 85–109.

Smyth, S., Kerr, W. and Phillips, P. (2017) Labeling demands, coexistence and the challenges for trade. *Journal of Agricultural & Food Industrial Organization* 15(1), 1–10. DOI: 10.1515/jafio-2016-0015.

# 9 Genetic Diversity in the Agro-Industrial Era: What Is at Risk?

*Each of us has a different set of life experiences with agriculture. Farming was the main profession of our ancestors, both in Canada and in the British Isles, where our families lived before emigrating to North America. Our immediate personal experiences, however, vary. Stuart, our youngest collaborator, was born into a farm family in the wheat belt in south-eastern Saskatchewan, was raised on a farm and spent part of his youth tending animals and cultivating crops. Peter, in contrast, is now the second generation on both sides of his family born off-farm. His paternal grandfather proved up a homestead on the bald prairies but let it lapse as he lost enthusiasm for the lonely existence of farming. His maternal grandparents were homesteaders in the Porcupine Soldier Settlement in the forest-fringe area in north-eastern Saskatchewan, carving out a profitable and productive farm before retiring to the west coast. While not a child of the farm, Peter's father, Bob, spent most of his career working with a large grain cooperative, first as a policy advisor and then running a farm weekly newspaper and reporting on the changes emerging in the sector. Bob always said his expertise was farmers and not farming – Peter has continued that mission. Bill's paternal great-grandfather was one of the first settlers in his area of Manitoba, leading a large family of 13 children to break the land for a homestead farm near what is now Hamiota. Tales of that hard life have become lore among the extended family that spread across the Canadian west. The rangelands and ranches of the Cariboo country of British Columbia played large in Bill's youth and continue their influence to this day in his passion for rodeo and the formalized ranching skills upon which the sport is based.*

*Given this background, we are an odd group of scholars to have engaged with one of the most compelling narratives in the modern agro-industrial system: the question of if or how modern agriculture supports or threatens traditional farming and the stock of global genetic resources created over the millennia by farmers and their communities. It would be fair to say that each of our family's farms ran commercial ventures, adopting the best-available technologies, seeds and breeds in an effort to realize a sustainable operation. The longevity of those ventures, both in our families and with their successors, suggests that something must have gone right.*

*Both individually and collectively, however, we have been drawn into study and debate about the balance between the commercial agro-industrial model that dominates in advanced industrial democracies and the more traditional cultivation of plants and animals using age-old techniques, knowledge, landrace seeds and animals to eke out what is too often today mere subsistence in various food-insecure parts of the world. This chapter explores our collective experiences with trying to reconcile our focus on profitable agro-industrial activities with our agreement that there is some intrinsic and functional value in traditional practices and genetic resources.*

## Farm Diversity in the Global System

At one time, almost everyone in the world engaged in farming or depended directly for subsistence on the activities of family members who were farmers. Even today, one in three workers in the world earns their primary income from farming or in industries directly connected to farming. In the mists of time, our ancestors developed basic tools, including axes, hammers, blades and ploughs, and used those tools to begin to meld nature to their needs. Historian Jared Diamond tells compelling stories about how hunter-gatherer societies would have observed that plants and animals grow at differential rates and provide varying nutritional value, flavour and convenience. Behavioural scientists today affirm that one of the distinct characteristics of humans is our ability to recognize and respond to patterns, real or imagined. While there were undoubtedly missteps, over time, humans began to recognize the value of spatially based resources such as grasses, fruits, berries, grazing animals and insects. Even without any self-aware grand plan, the selective pressures of individuals and communities gathering and culling plants and animals would contribute both to refinement of our tastes and of our food sources. Most anthropological histories, based on evidence found in archaeological digs, assert that humankind began to settle as purposeful cultivation of plants and animals took hold. In the first instance, as communities recognized the mobility of animals, they would seek to settle in set places at optimal times for hunting, while those who found use in various plants began to recognize the value of first returning to the most productive sites and then protecting those sites from animal, insect or human predators. Scholars think that the simple nutritional pay-offs to individuals to settle and exploit local supplies over time led to simple selection of both our first identifiable cultivars and the domestication of ruminants. Once people settled in communities, they needed new forms of organization to sustain their health and security. Farming thus became inextricably interwoven with civilization.

Diamond and others note that the efforts of farmers in the temperate zone, mostly between 20–40° latitude, in both the northern and southern hemispheres, were the foundation of the global food system today. Most of the main food crops and almost all of the animal species that make up our diets were first domesticated in these geographic ranges and then, through migrations and trade, were distributed globally. Scientists and historians have identified that between 5000 and 10,000 years ago, people in 'centres of origin' created the germplasm base for our main sources of plant protein today, including wheat (*Triticum* spp.) in south-western Asia, rice (*Oryza sativa*) in China, maize (*Zea mays*) in Mesoamerica, potato (*Solanum tuberosum*) in the

Andes and Amazonia, sorghum (*Sorghum* spp.) in the Sahel, oil palm (*Elaeis guineensis*) in tropical West Africa and sugar cane (*Saccharum officinarum*) and banana (*Musa sapientum*) in New Guinea. Meanwhile, sheep and cattle were domesticated first in western Asia, pigs in the Middle East and chickens in the Far East. Migration and diffusion of these key crops and animals was facilitated by the common climatic conditions in the temperate zone.

As civilizations emerged and began to trade and go to war to extend their power, influence and wealth, genetic materials migrated with the victors, either to conquered lands to secure influence or within the expanding empires, in order to exploit the best of what the victors had acquired.

The age of empires signalled the first tipping point in our story. Before about 1500, most of the migrations and conquests were episodic, random or opportunistic, or at least mostly limited to continental moves. By the 1500s, however, a number of political, economic and technological events converged to generate a race by the big five European empires to expand. By the 16th century Europe had been carved up into a number of growing or nascent empires – initially Portugal and Spain, followed by England, France and Holland – which needed both more food to support rapid growth in rapidly urbanizing populations and new markets to replace the overland routes to Asia, which had been cut off with the fall of Constantinople in 1453. Meanwhile, innovations in ship designs, cartography and navigation enabled ships to go farther afield, affording new discoveries. Might was undoubtedly right in this period, but the claims of the powerful were afforded cover with the papal bull *Inter caetera*, issued by Pope Alexander VI on 4 May 1493, which effectively divided South America between Spain and Portugal and excluded the Spanish from the route around Africa to Asia in favour of the Portuguese. The papal bull decreed that 'the Christian religion be exalted and be everywhere increased and spread, that the health of souls be cared for and that barbarous nations be overthrown and brought to the faith itself'. In effect, Christian countries now had a duty to proselytize through conquest. Any territory occupied by pre-contact people or those who had not converted to Christianity were deemed 'empty'. This set up one theme of the debate today – many societies reject the Christian hubris of this approach.

More importantly, conquest was purposeful. The goal was to increase the volume and value of resources under the victors' control. Beyond the obvious value of access to mineral riches, plunder and slave labour, the genetic resources of those disparate parts of the empires offered real long-term value. The flow of material was, however, two-way. Cereals migrated throughout the temperate zone as migrations and conquest continued, rice diffused across the world and potatoes and maize were transferred to Europe and beyond. There were also transfers of plants within more tropical areas. The British ship HMS *Bounty* (of mutiny fame), for example, was on a voyage to acquire and transfer breadfruit plants from the South Pacific to the Caribbean to provide a dietary staple in the West Indies.

These transfers between and within large continental ecosystems transformed the landscape. Ever since, we have experienced a collective frisson as we contemplate the fundamental changes we have made to the natural environment. Modern agriculture, with its greater uniformity and stability of genetic materials, has delivered a cornucopia of food. However, those with an interest in and focus on centres of origin and centres of genetic diversity worry that we have lost much in our search to create greater utility and function in the narrow range of plants and animals that make up the bulk of our diet. Some of our debate today is about whether or how the world in general, and the advanced industrial economies in particular, should acknowledge and compensate the countries or centres of origin for these profoundly useful foods that were appropriated and disseminated globally.

International efforts to balance these two perspectives have varied widely. We have spent most of the past 100 years in *ex situ* efforts to link and secure collections of genetic materials in national and international gene banks and the Global Crop Diversity Trust (GCDT) seed vault, and in assisting with *in situ* (on-the-ground) farmer-based conservation programmes. We have also invested significant energy to develop a set of rules, processes and agreements to acknowledge, manage, access and in some ways monetize and compensate traditional communities for their genetic assets. As will be discussed further below, all these works in progress are laudable but may be overtaken by the emergence of a new set of digitally rich breeding technologies that appear able to disembody the functional value of much of our genetic diversity from living specimens.

## What Are Plant Genetic Resources and Why Should Anyone Care?

Plant genetic resources (PGRs) have meaning at several levels, which varies depending on whether one is interested in the practical economic application in agriculture, or the symbolic or cultural value to society.

The FAO's *Glossary of Biotechnology for Food and Agriculture* defines PGRs as the reproductive or vegetative propagating material of: cultivated varieties (cultivars) in current use and newly developed varieties; obsolete cultivars; primitive cultivars (landraces); wild and weed species; near relatives of cultivated varieties; and special genetic stocks (including elite and current breeders' lines and mutants). At one level, PGRs are based on the original natural distribution of genetic materials before the intervention of man. Over the intervening millennia, farmers, and more recently organized public and private plant-breeding programmes, have dramatically altered the phenotypes of these plants, to the point where they would be unrecognisable to the original developers. Along the way, these plants were introduced and adapted to new regions, hence most of the world's production of these crops is now in areas far removed from the centres of origin. The accumulated genetic resources of the collective effort of mankind are the subject of the debate outlined below.

In 1996, the FAO issued its first *State of the World's Plant Genetic Resources*, warning of a large-scale loss. The report argued that the spread of modern, commercial agriculture and the introduction of new varieties of crops contributed to the loss of genetic diversity. The erosion of biodiversity and the loss of genes in those species was viewed as a threat to agricultural and food security, especially for the abjectly poor who rely on biological products for 85–90% of their livelihood needs (i.e. food, medicine, fuel, fibre, clothing, shelter, energy and transportation). While loss of biodiversity pre-dates the advent of biotechnology and private intellectual property rights (IPRs), both developments are accused of exacerbating the problem.

No one knows exactly how much diversity once existed in domesticated species, so it is impossible to say exactly how much has been lost. One study estimates that 34,000 species of plants – estimated to be about 12.5% of the world's current flora – face extinction. The 1996 FAO report on PGRs cited some examples of the loss of biodiversity. In China, of the nearly 10,000 wheat varieties in use in 1949, only 1000 remained in the 1970s, while in the USA, 95% of the cabbage (*Brassica oleracea var. capitata*), 91% of the field maize (*Zea mays*), 94% of the pea (*Pisum sativum*) and 81% of the tomato (*Lycopersicon esculentum*) varieties cultivated in the 19th century have been lost. Many other species are suffering a similar fate in Europe, Africa, Asia and South America. How much is lost in natural systems is largely unknown.

Loss of biodiversity poses two main concerns. First, the concomitant increase in uniformity of plant genetic material can lead to greater risk and uncertainty. Considerable genetic uniformity now exists in crops such as hybrids of rice and sunflowers (*Helianthus* spp.) and in a few crops like canola (*Brassica* spp.), which can be traced to a single genetic donor more than 60 years ago. The FAO notes that uniformity in the rootstock of California wine grapes (*Vitis vinifera*) and the resulting susceptibility to virulent disease forced wine producers to dig up and replace their vines at the cost of hundreds of millions of dollars. Second, it has been argued that for every plant that becomes extinct, 30 other species go with it, many of them microorganisms. This is viewed with concern by some biologists because many plant pathogens (including fungi, viruses and bacteria) are believed to play a vital role in the functioning of ecosystems, seed function and soil–root interactions. While commercial agricultural practices are not the only reason for loss of biodiversity – degradation and destruction of forests and bush lands, overgrazing, wars and civil strife also threaten ecosystems – the international agricultural research community believes this is one area that they have the potential to influence.

One solution has been to develop *ex situ* gene banks to house extensive collections of landraces and other varieties. The FAO estimates that there are about 7 million accessions being stored worldwide in more than 1300 gene banks, in 77 countries. There is real concern that the infrastructure supporting those collections is at risk. The FAO reports that equipment problems (particularly cooling and humidity control), insecure electrical supply, difficulties in seed drying (especially in humid regions) and inadequate investment in seed regeneration is endangering seed quality. The FAO has stressed that seed conservation in gene banks (*ex situ*) and in ecosystems and natural habitats (*in situ*) are complementary, and both should be strengthened. The

*International Treaty on Plant Genetic Resources for Food and Agriculture*, through the GCDT, since 2003 has sought to raise an endowment to fund in perpetuity the 20 international collections of 17 major food crops in nine gene banks of the CGIAR, the Svalbard Global Seed Vault and the Nordic Genetic Resource Centre. As of 31 December 2017, the GCDT had pledges for US$440 million and had plans to raise that to US$850 million.

Many people and societies vest much more in their plants than the simple genetic and agronomic vitality of the seeds. Plants can play a major role in the economic, social, cultural and historical life of traditional communities, and thus citizens are concerned about how the plants they have developed and nurtured are viewed and used by others. The Crucible Group argues there is an important link between PGRs and cultural vitality. They asserted that local and Indigenous peoples who speak ancestral languages are severely threatened by loss of sovereignty over land, resources and cultural traditions, which tends to be linked to control over their indigenous PGRs. As communities become increasingly marginalized, local people lose local agronomic knowledge, innovative capacity and wisdom about species and ecosystem management. Ultimately, there is concern that the reductionist approach of modern science, and biotechnology in particular, does not respect the vast body of traditional knowledge about plants and their place in the ecosystem built up over the millennia.

The debate about PGRs has largely been precipitated by the extension of private IPRs. While there is a long history of countries trying to protect their indigenous plants – India in the 1700s, for instance, had a penalty of death for anyone illegally exporting rice for propagation while plantation agriculture was a common model in many countries – it was really only in the 20th century that these rules were adopted by states more generally to explicitly encourage investment in crop invention and innovation.

By the 1930s, the US patent system granted rights for asexually reproduced plants (e.g. by cuttings and grafts) and in the 1940s a number of European nations granted plant-breeders' rights for new varieties developed domestically. Plant-breeders' rights were extended internationally in the 1970s through negotiation of the International Union for the Protection of New Varieties of Plants (known as UPOV based on the French name). During the 1980s, rights were revised and extended as new biotechnologies enabled new inventions. By 1985, the US Patent and Trademark Office had granted its first patent on a whole plant. This effort was internationalized in the 1990s, with the negotiation of the WTO in 1994 and its Agreement on Trade-Related Aspects of Intellectual Property Rights (TRIPS), which requires all member states (165 in 2018, collectively representing more than 96% of global production and trade) to either grant full patent protection to plants or to provide a *sui generis* (i.e. purpose-built) system of protection for plant inventions. All developed countries had until 2001 to comply with TRIPS, while developing countries had until 2006 to comply.

The extension of private rights to PGRs has divided the research and farming community, creating both owner and user classes. Formal IPR structures – including patents, plant-breeders' rights and trademarks – are all designed to grant a limited monopoly (by time, application or market) for the 'fruit of one's labours' in exchange for full disclosure of how the invention works. The various mechanisms are designed to provide a right to exclude others from the use of an invention for a limited term in order to allow the inventor or creator to use it and recoup their investment. All these mechanisms require a clearly identified inventor or creator, a clearly defined inventive step (that is new and not obvious), an intended use and full disclosure of the method. In that sense, they are not able to protect or support collective or public efforts that happen over a long period of time.

Private IPRs have been deliberately adopted in order to encourage greater private, for-profit research and to encourage more rapid commercialization of inventions. While many microeconomic studies indicate that in some ways the extension of property rights has accelerated investment, invention and commercialization, some scientists are concerned that extension of private rights to fundamental processes, specific genes or whole organisms could block other users.

Some farmers and communities are concerned that their role in creating and conserving PGRs, especially landraces and the related traditional knowledge, is being or will be disadvantaged by our current system. Some worry that they may lose legal access to PGRs that they themselves developed and conserved if they become included in a patented variety. There have been a few high-profile cases in which the 'inventions' in property claims are simply rediscoveries or even thefts of traditional knowledge. At root are questions of access, benefits sharing, ownership and control. A related debate centres on whether private IPRs are the only or the best method of supporting and encouraging research into plants for non-industrialized farmers. Other strategies that might be more compatible with traditional PGRs include more public investment though consortia, public research programmes or subsidies to private companies to undertake directed research.

## Genetic Pollution? GM Maize in Mexico

One of the subtexts to the debate about coexistence is whether and how modern, improved varieties embodying GM traits can coexist with historical varieties and traditional knowledge, especially in centres of origin and biological diversity. This issue came to a head in 2001, when David Quist and Ignacio Chapela, two University of California Berkeley plant biologists, published in *Nature* that they had detected transgenes in local maize landrace varieties in the mountains of Oaxaca, Mexico. A pressing concern for many was that none of the transgenes found had been approved for deliberate environmental release in Mexico, and that if left unchecked, the transgenes might degrade maize genetic diversity. The resulting flurry of activity engaged local farmers (*campesinos*), a host of environmental NGOs, the main biotechnology companies that had developed the disputed traits, a range of scientists around the world, various state authorities and the federal government in Mexico and, ultimately, a reference to the North

American Commission for Environmental Cooperation, the special agency that emerged from revisions to the NAFTA in 1994. Peter ended up as one of three Canadians on the Chapter 13 Panel (a process for concerned governments or groups to challenge domestic policy measures), which was tasked with reviewing the issue and reporting to the three national governments on their findings and recommendations.

Mexico is regarded as the birthplace of cultivated maize. The leading hypothesis is that maize descended from teosinte, a grassy plant common in the highlands in Mexico. Dozens of local landraces of maize have emerged and been cultivated for between 5000 and 8000 years; the genetic base of these landraces varies widely by location and over time. Wild teosinte populations persist today, often as part of the agro-ecosystem where commercial maize is grown, and because they are the same species, gene flow among the two crops has been observed.

Maize in Mexico is the most important crop cultivar, with 7.9 million ha of production in 2001. Most domestic production is white-maize varieties for human consumption. The country also imports more than 5 million t annually, largely yellow maize for animal feed. Maize varieties currently cultivated include modern, improved, hybrid varieties developed by multinational seed companies, open-pollinated, improved varieties developed by national public research institutions and small companies, and landraces conserved and cultivated by *campesinos*. Although there was evidence at the time of the case that new hybrids of maize introduced into Mexico could and did contribute genes to local landraces, the idea of introducing transgenes from unrelated species into the landraces and wild relatives of maize was a matter of deep concern both in Mexico and internationally.

Globally, maize (called corn in North America) has been the target of aggressive genetic transformation for more than 90 years, from the invention of hybrid maize in the 1920s, through the widespread adoption of hybrid maize in the post-war era, and into the Green Revolution, where herbicide and fertilizer innovations for wheat and rice were translated to maize. Maize was also an early and important target for GM crops. By 1988 transgenic maize crops were in confined field trials in the USA and a range of other countries. The first commercial GM maize varieties were approved for unconfined production in the USA and Canada in 1995 and entered the export market in 1996. Since then, more than 230 genetic traits (called events by the industry and regulators) have been introduced into about 10 food crops; most involve insect resistance and herbicide tolerance. While some of these candidate traits were introduced into maize and grown in confined field trials in Mexico, none of the varieties have yet been approved for unconfined cultivation in that country, at least partly because of concerns about the effect of these crops on landraces and the indigenous agro-ecological system operating in many regions of the country.

In the controversy that erupted in 2001, Quist and Chapela reported that they had used a range of tests that detected the presence of the specific genetic promoter which was then used in all GM maize varieties commercially available in the USA. Once the results were released in the journal *Nature*, a host of critics emerged to challenge or debate the findings. In the first instance, a broad section of the scientific community commented on the findings and wrote to *Nature* and other journals disputing the technical results. They argued that the tools used were especially prone to produce false positives. *Nature* received numerous communications about the article and its findings, and the editors corresponded with the authors in search of further evidence, but in the end the editor took the unusual step of publicly reporting that the evidence in the article was not sufficient to justify publication of the paper.

Many in the scientific and industrial community hoped that the declaration of the *Nature* editor would be the end of the matter. However, the underlying issue remained unanswered. This report simply added to an already simmering debate. There had been early warnings of a high probability of transgenic maize entering Mexico as early as 1995. Greenpeace had reported in March 1999 that they had found evidence of transgenic maize in commodity shipments imported from the USA at the port of Veracruz. In response, the Mexican government commissioned a larger study involving the Centre for Research and Advanced Studies of the National Polytechnic Institute (a Mexican non-governmental scientific research establishment), and the Institute of Ecology at the National Autonomous University of Mexico. Further tests conducted in 2001 in 188 communes in the Oaxaca region revealed that more than 7% of the maize plants examined contained genes from GM maize; in subsequent investigations in 2002 the reported incidence fell to 0.11%. Other field studies were undertaken in an effort to confirm or refute the results. The first peer-reviewed study (in 2005) found no evidence of any transgenes in the crop or in government storage facilities in either the 2003 or 2004 crop years. Over time, a range of scholars and practitioners came to a somewhat unsatisfactory scientific judgement that gene transfer (technically described as introgression) was both likely and inevitable given the context for the traditional agro-economy in Mexico but that the biological impact on both farmer production and biodiversity would be minimal or undetectable in the long term.

Nevertheless, public controversy had continued to build, both in Mexico and throughout a segment of the global NGO community. In April 2002, the Commission for Environmental Cooperation (CEC), a creation of the NAFTA side deals in 1994, was petitioned – by 21 Indigenous communities of Oaxaca, three Mexican environmental groups (Greenpeace México, the Mexican Centre for Environmental Law and the Union of Mexican Environmental Groups), and eventually supported by more than 90 letters from organizations and institutions throughout the three NAFTA countries – to undertake an investigation. This issue was considered of great potential environmental importance, given that Mexico is a centre of origin and diversity for maize and that maize is so intrinsically linked to Mexican culture, especially that of Mexican Indigenous groups. An Advisory Group on Maize and Biodiversity, involving 16 participants from industry, academia and NGOs in Canada, the USA, Mexico and the UK, was commissioned to use objective science and the participation of stakeholders to transparently develop an advisory report. The advisory group, in collaboration with the CEC secretariat, commissioned a set of background chapters from a range of academics and practitioners about specific topics related to maize. The papers were assessed by the group, peer reviewed and published in a compendium to the final report. The secretariat and the advisory group also convened an open public symposium in Oaxaca, at which summaries of the chapters were presented and discussed. The advisory group then met and formulated a set of key findings and recommendations based on the background chapters and the peer reviews, the professional expertise of the members themselves, comments received at the public symposium and, subsequently, the comments from the governments of Mexico, the USA and Canada.

The advisory group accepted that maize had significant cultural, symbolic and spiritual value for most Mexicans and that many saw transgenic gene flow as a direct threat to political autonomy, cultural identity and biodiversity. The issue of transgenic maize became entwined with historical issues and grievances affecting rural Mexicans that are not directly associated with either improved maize or traditional landraces: Indigenous peoples have a history of perceived inequity and injustice at the hands of Mexicans of Spanish origin, Americans and powerful elites, which has led to high levels of poverty, food insecurity and dislocation.

The advisory group learned that an estimated 84 ethnic groups in Mexico have been long-term stewards of 59 identified Mexican maize landraces. In pre-Columbian times, these ethnic groups occupied the whole country, but they retreated to the sierras during European conquest. Geographic isolation generated a range of productive ecosystems. Experimental planting and breeding of maize was identified as a millennia-long tradition. Mexican landraces are neither genetically static nor genetically similar: they are constantly being changed by those who use them. As part of this process, genes from improved, modern varieties are inadvertently, sometimes deliberately, introduced into the landraces. The CEC advisory group heard evidence that *campesinos* regularly test materials drawn from government food-aid channels or from the commercial seed trade and cross them with their traditional cultivars. The evidence strongly suggested that the observed gene transfers were probably a result of *campesinos* either inadvertently or intentionally planting transgenic maize and crossing those traits into landraces, at least partly because there was no visible way of differentiating transgenic from non-transgenic seed.

On the core issue of genetic diversity, the panel concluded that gene flow would occur but whether it would persist in plant populations depended on whether the gene flow was a one-time or recurrent event, the rate of gene flow, and whether the new genetic material was detrimental, beneficial or neutral to the host plants. The group concluded there was no reason to expect a transgene would have more impact on the genetic diversity than other genes crossing from modern cultivars. The maize gene pool represents tens of thousands of genes, with significant variation within and among populations, so that the transfer of a few genes was unlikely to have any major effect on genetic diversity.

## The International Effort to Govern Genetic Resources

Over this last 50 years, a large number of governments, institutions and organizations have been involved in addressing concerns about PGRs, including access and benefits sharing (ABS). In some respects, there are both too many actors and not enough effort. Although individual countries have attempted to control access to their genetic resources for hundreds of years, recently there has been a diffusion of effort. Now we have an alphabet soup of international agencies, treaties and committees which have produced a wide range of proposals, guidelines and commitments, but little in the way of operational or judicable structure.

Recently there has been significant debate and effort invested in negotiating a range of international conventions or treaties to delimit and protect indigenous rights to genetic resources, involving the International Labour Organization, the United Nations (UN) and Inter-American declarations on the rights of Indigenous peoples, the UN Development Programme, the UN Conference on Trade and Development and the European, Asian and African Development Banks. In the context of PGRs, in particular, the special institutions involved in delimiting rights and facilitating ABS include the Convention on Biological Diversity (CBD) (1992), the Bonn Guidelines on Access to Genetic Resources and Fair and Equitable Sharing of the Benefits Arising out of their Utilization (1983), Agenda 21 and the Cartagena Protocol on Biosafety (2000), the International Undertaking on Plant Genetic Resources (1983), the International Treaty on Plant Genetic Resources for Food and Agriculture (2001), the CGIAR and related gene banks and various national programmes (e.g. the Canadian and Swedish International Development Agencies). The Nagoya Protocol on Access to Genetic Resources and the Fair and Equitable Sharing of Benefits Arising from their Utilization (2014) is a supplementary agreement to the CBD directly tasked with this issue. Protocol members seek to advance the policy and practice of accessing and utilizing genetic resources and associated traditional knowledge in research and development activities. The protocol created a clearing-house mechanism to facilitate the collection and sharing of information on the national implementation of ABS, including legislative and other measures, permits, relevant authorities and institutions as well as codes of conduct and best practices.

What may not be obvious to the casual observer is that there was a sea change in approaches between the 1983 International Undertaking, which defined PGR as the 'common heritage of mankind' and the 2001 International Treaty, which reaffirmed or reassigned rights to nation states and their Indigenous communities. Since then, many countries of origin and countries of diversity have moved to develop national laws or have worked with neighbouring countries to develop regional structures and rules to govern their property claims.

Meanwhile, the commitment of some key agri-food producing and exporting nation states and much of the research establishment is variable, at best. The USA, for example, is not a party to the CBD and hence is not bound by its provisions. Other key countries, such as Canada, are signatories to the CBD but have declined to fully engage or commit to hard measures that would entrench rights to traditional knowledge and ABS. The EU as a whole supports the effort to strengthen the system but many of its member states are less committed. Even countries representing centres of origin or diversity have differing approaches and perspectives and seldom are willing to combine their efforts to advance the international agenda.

In effect, the world has created a relatively dense system of national and international rights, claims and processes for handling disputes, that all rely on a body of evidence to adjudicate. While these efforts may yield some value, ultimately what is protected and how it is protected and used is now mostly a matter for nation states to decide domestically, which renders international efforts to benchmark norms and standardize benefit-sharing requirements difficult.

## The Digital Future and ABS: Where to Now?

As scholars engaged with the biosciences, we have been watching and studying the impact of automated sensing and advanced data analytics on both the production and logistics system and on the research space itself. The rapid convergence of biology and cyberspace has created a host of novel opportunities as well as challenges. The economics of genomics has significantly accentuated this process. The cost of sequencing a genome decreased from US$100 million in 2001, to approximately US$1000 in 2017. The cost per raw megabase of DNA sequence (a million bases) declined from US$10,000 in 2001, to less than 1 cent in 2017. The genome of an organism contains its genetic information (or code), which is responsible for the development and homeostasis of organelles, cells, tissues, organs and organ systems. This information can now be uncovered and subsequently stored, edited and transferred digitally. Perhaps most importantly, it can be reconverted into tangible biological constituents through gene editing. Scientists seldom focus on a specific specimen to sequence; instead they sequence all or part of thousands of plant samples originating from various sources. This process makes it operationally very difficult to track the depth and scope of information utilized from any individual source in a final product. Furthermore, digital technologies permit storing, distributing and analysing the disembodied data with low or zero marginal costs. Novel genome editing techniques, big data and synthetic biology to varying degrees have begun making use of digital sequencing information (DSI). Individually and collectively, because of their transformative capacities, these fields hold great potential to address issues in crop breeding, energy production and health.

This excites interest on a variety of fronts. In the first instance, international negotiators at the Nagoya Protocol have begun discussions about what, if anything, they can do about this. It is not yet clear whether any progress can be made. Complicating this is the reality that the USA is the major site for integrating digital databases with new breeding technologies, but as with most policy conversations about ABS, the USA does not have legal standing in the forum. A few national governments and advocacy groups have mused and tested to see which of the modern property-protection mechanisms might work to sustain some control over the disembodied genetic materials derived from PGRs. Some have asserted copyrights or patents might work, in the same way computer algorithms can at times make use of those mechanisms. To date, however, none of the options shows much promise for having much more than limited effect.

Meanwhile, bench scientists are excited by the prospects of using the digital databases and new breeding techniques to reintroduce or expand the genetic diversity within elite breeding lines in the major food crops, for the first time bringing modern science to support the vitality and security of many of the secondary and niche crops that are most important to subsistence farmers in food-insecure parts of the world. No doubt part of that enthusiasm is motivating some of the recent restructuring of the major multinational biotechnology industry, but it doesn't stop there. Public breeders in CGIAR centres, in universities and in national agricultural research stations all see potential to codify and secure our knowledge of PGRs and genetic diversity through the application of gene sequencing and genomics, genome-wide associations studies and a range of other technologies that allow for greater specificity of the structure and function of aspects of a species.

In this context, Peter was drawn into an international public effort to help manage this transformation. DivSeek, an international consortium of public breeders, scientists and gene banks, seeks to define and create the information management architecture to support the application and use of DSI, either for use with conventional or new plant-breeding techniques. The mission of DivSeek is to 'cross-link, support and add value to individual activities that harness the power of crop diversity for food and nutritional security and societal and economic benefits, by enabling breeders and researchers to mobilize genetic variation in order to accelerate crop improvement'. As of Autumn 2018, DivSeek had 68 partners that were engaged in the venture's efforts to formulate and advocate for common data and informatics standards and best practices designed to enable interoperability among information systems, all in an effort to link germplasm with the passport, characterization and evaluation data that is connected to materials in gene banks. DivSeek has working groups on data standards for interoperable tools, including Global Open Data for Agriculture and Nutrition (GODAN); gene bank information management systems; genomics for gene banks; phenotyping for seed diversity; semantics for harmonizing trait and agronomic data (e.g. ontologies and concordances); translational approaches for minor and underutilized crops; and a FAIR Data project. The goal is to publish existing and new datasets so they can talk to each other (what is formally described as a semantically interoperable format). By appropriately annotating data items and metadata, computer systems should be able to automatically or semi-automatically combine and analyse different data sources, resulting in richer knowledge discovery. If successful, this public consortium offers the potential for digital innovations to support and sustain PGRs that accelerate both private research and public benefits.

As with many complicated and socially embedded controversies, there are no simple fixes that resolve matters completely. What PGRs are, how they have been, are and will be used, and who has legitimacy and capacity to define terms of use or to share in the benefits, remain open questions. This frustrates many Indigenous communities and their advocacy communities, generates uncertainty in the scientific and business communities and confounds governments. Based on our experience, the best we can probably expect over the coming decades is more muddling through, with individuals and institutions whittling away at those conflicts that truly block constructive development and use of global PGRs, and addressing any resulting egregious inequities.

## Further Reading

Bubela, T. and Gold, E.R. (2012) *Genetic Resources and Traditional Knowledge: Case Studies and Conflicting Interests*. Edward Elgar, Cheltenham, UK.

Crucible Group (2000) *Seeding Solutions Volume 1: Policy options for genetic resources: people, plants, and patents revisited*. Co-published by the International Development Research Centre, the International Plant Genetic Resources Institute and the Dag Hammarskjold Foundation. Available at: https://www.idrc.ca/sites/default/files/openebooks/316-x/index.html (accessed 12 February 2019).

Diamond, J. (1997) *Guns, Germs and Steel: the Fates of Human Societies*. WW Norton, New York.

FAO (1996) *State of the World's Plant Genetic Resources for Food and Agriculture*. FAO, Rome. Available at: http://www.fao.org/agriculture/crops/thematic-sitemap/theme/seeds-pgr/sow/en/ (accessed 12 February 2019).

Oguamanam, C. (2018) *Genetic Resources, Justice and Reconciliation: Canada and Global Access and Benefit Sharing*. Cambridge University Press, Cambridge.

Phillips, P. and Onwuekwe, C. (2007) *Accessing and Sharing the Benefits of the Genomics Revolution*. Springer, New York.

# 10 The Sustainability Challenge: Are We Facing a Chemical Armageddon?

*For most of us, much of our knowledge of the chemical use in agriculture comes from our vegetable and flower gardens. Every spring as the snow recedes in my back yard, my work shifts from shovelling snow to raking dead grass, trimming dead raspberry canes and tilling my garden. There is something very pleasant about the smell of freshly tilled soil that clearly must resonate with growing up on a farm. Not that I can say it was terribly enjoyable in my youth on my family farm, spending long, hot summer days driving back and forth across a summerfallow field with a 12-ft-deep tillage cultivator pulled by an open-air Case 730. Summerfallow is the term used to refer to when a field is not planted to a crop for an entire growing season but is tilled on a regular basis to kill the weeds that have germinated since the previous tilling. The reason the smell of that freshly tilled soil stands out for me is probably that our summerfallow had to be worked roughly every two weeks or about 10 times a summer and the inside of my nostrils were frequently coated in dust, preventing any sense of pleasure.*

*With my back yard garden tilled, I plant a variety of vegetables and then spread some nitrogen fertilizer over the entire space. For the most part, I limit the chemical use in my garden to what is essential. I have to spray my raspberries with a chemical that controls insects, as a few years ago I started to discover small worms in the ripe fruit. Nothing is more off-putting than sitting down to enjoy a bowl of fresh raspberries and seeing a small white worm wiggling its way out of one of the berries. On a few occasions, I've had to spray my potatoes to control potato beetles. I've also learned the hard way how cabbage moths will also lay eggs on broccoli, making detection very difficult. Small green worms on a green vegetable raises the likelihood of worm consumption.*

*Now, I'm not the kind of gardener that runs around misting things with one chemical or another every other day of the week, but I have learned that the occasional and controlled use of chemicals ensures that I'm able to enjoy the fruits (and vegetables) of my labour. A couple of years ago, my neighbours had to remove three apple trees and a lovely crab apple tree as insects got into the trees, destroying all the fruit as soon as it started to ripen. It was a pleasure picking a few pails of crab apples and making jelly, however seeing the fruit ruined by the presence of worms inside*

S.J. Smyth *et al.* (2019). *GM Agriculture and Food Security: Fears and Facts* (S.J. Smyth *et al.*)

*virtually every apple or crab apple was sad. Losing four lovely trees from their back yard also left a gaping hole in the natural shade provided to their house and deck.*

*Insects don't only attack fruits and vegetables, they are also fond of many varieties of flowers that get planted in flowerbeds. It is frustrating to see a rose bush virtually defoliated by insects in the matter of a few days, often killing the rosebush. Sure, this could be less of an issue in warmer climates where roses can easily be replaced, but in the colder northern climates, where considerable preparation is spent in the fall ensuring that the rose is protected to the best of one's ability to survive the frigid winter temperatures, it is a great joy to see a rose begin to leaf out in the spring. It is a sign of the hardiness of the plants around us, which make the loss to insects even more difficult to bear.*

*Nature is ferocious. Weeds are pervasive and hardy while insects of one sort or another relentlessly attack the fruits, flowers and vegetables commonly found in millions of urban yards. The simple reality is that while a person may not like having to spray plants in their yard with chemicals, if one wants to enjoy the rainbow of colours offered by flowers, or the fresh fruits and veggies provided by back yards and garden spaces, chemicals are essential to ensuring success.*

*Farmers are no different. They too want to improve their chances of success, only in their case failure can lead to losses that dwarf any frustrations I face in my small garden. The spraying of crops for weeds or insects varies dramatically between the large farms in industrialized countries and the small landholder farms in developing counties. When a farmer in Canada or the USA pulls into a field to spray it, they do so with a high-clearance, motorized sprayer, that has a clearance of about 6 ft and that has spray booms that can span 130 ft. Protective suits and ventilation masks are worn while pouring chemicals into the tank to minimize inhalation or skin contact. A water truck is also present to pump the required water into the sprayer's tank; these tanks can be as large as 1600 gallons, the equivalent of filling a Honda Civic gas tank 130 times. The farmer sits in a sealed, air-conditioned cab, with GPS steering that minimizes the overlap between passes to one-quarter of an inch, which is roughly the width of a pencil.*

*Compare this to the spraying of a small field in a developing country. Cotton is a crop that must be sprayed multiple times per growing season to prevent insects from damaging the cotton bolls, which would reduce the quality and revenue for the farmer. Cotton farmers in the past in China and India would typically spray their cotton fields 25–30 times per season. The water would have come from a community well and the chemical would have been hand poured into the spraying equipment by the operator, wearing no protective clothing. The cotton plant would be about 3 ft high when sprayed and this would be done by an individual (often a teenage child) walking between the rows of cotton plants, wearing a backpack sprayer with the wand in their hand. The person spraying would commonly wear a t-shirt, shorts and sandals. To spray the rows, they would walk forward, moving the wand back and forth to spray the plants in the row. As they walked down the row, the leaves of the plants would brush against their bare arms and legs, resulting in some of the chemical coming into contact with their skin. When they finished spraying all the rows in the field, they would take the backpack sprayer off, but there would be no, or very limited, opportunity for them to wash the chemical residue off their arms and legs, so it would be absorbed into the skin. Depending on the toxicity of the chemical being used and how rigorous the chemical regulations were in the specific country, the chemical could be rather benign to humans or have quite an adverse effect, causing what is known as pesticide poisoning.*

*This is the reality of agriculture today – that new technologies can help to increase the sustainability of our food system, but we have a variety of challenges and opportunities that require different care and attention.*

## The Historical Role of Chemicals in Agriculture

For millennia, farmers have undertaken efforts of every nature to try to protect their crops against threats that could reduce their yields. Throughout history, field sizes were small and farmers, who in real terms were often indentured to the local land owner, would spend all day in the fields hoeing and weeding by hand. The Crop Protection Research Institute has estimated that in the USA alone it would take about 70 million more workers, or more than 55% of the available workers, to hand weed the commercial crops cultivated today; currently only about 2 million people work in farming in the USA. The removal of weeds allows soil nutrients and moisture to be reserved for growing the crop that is planted. Healthy and competitive plants are also less likely to be targeted by pests. In the past, without effective weed control or synthetic insecticides, 'smudges' were often lit around the edges of fields, with the hope that the smoke would discourage the presence of insects. There is no evidence about the success of this technology as a means of controlling insects in fields and it is seldom used today, even where other options are lacking.

Over the course of history, while farmers have been physically capable of trying to ensure that weeds reduced their yields by as little as possible, they were largely at the hands of nature when it came to dealing with insects and plant diseases. Frequently, the results would be devastating. The Irish Potato Famine is one of the iconic examples of the vulnerability of plants to pathogens and the profound effects of agronomic failure on an economy and society. Imagine for a moment what the settlement of Australia, Canada and the USA would look like, were it not for the disaster. Were it not for the devastating and prolonged outbreak of potato blight, the populations of all four countries could be considerably different today. In 1845–1849 a disease called late blight, which destroys leaves and edible roots (i.e. the tubers we eat) devastated the potato crop for five successive years. About half of Ireland's population of 8.4 million then depended almost exclusively on potatoes for their diet. An estimated 1 million people died of starvation or opportunistic diseases like typhus between 1845 and 1852, and another 1 million emigrated in the hope of improving their circumstances. The Irish population dropped by about 25% in those 5 years of crop failure. The migration continued until 1901, when the population of Ireland was less than half of what it had been before the famine. My ancestors were part of this movement, leaving County Wicklow in the 1840s and immigrating to Southern Ontario in Canada.

In response to those kinds of challenges, chemistry has been developed and become a core part of commercial agriculture. From time immemorial, farmers have used animal and plant manures as fertilizer and learned to supplement their soils with additives, such as potash, to sustain and improve their yields. By the 1840s, just as plant breeding was starting to get organized, a range of inventors and industrialists in Europe were studying crop rotations and potential additives to crops to determine the optimal management of agriculture. Nitrogen-based fertilizers were beginning to be produced and applied by farmers and by the 1850s, several dozen chemicals were available for use. Often, the ability to apply the chemicals was an issue and it was not until the mechanization of agriculture in the middle decades of the 20th century that chemicals became commonplace in agriculture. The farmers I grew up with talked about first adopting chemicals in the 1950s as they gained access to tractors and sprayers which could be used to efficiently and evenly apply chemicals to larger fields.

Most farming systems around the world adopted chemicals aggressively after about 1950. A range of herbicides have been developed to kill weeds, synthetic insecticides to kill bugs, fungicides to control disease, fumigants and desiccants to assist with harvest and natural pesticides to control pests in organic production systems. In total, more than 4 million tonnes of active ingredients of pesticides were applied in 2016. About 70% of the active ingredients used in 2016, and the highest application rates per ha, were in South America and Asia, up 335% and 92% respectively since 1990. Chemical usage per acre has been rising in all major regions of the world, but rates of application are unambiguously too low in Africa, where weed, insect and fungal damage undercuts productivity, contributing to food instability.

Herbicides are king in the chemical trade. Herbicides make up 40–70% of total chemical usage in agriculture, depending on the year and market. The early adopters were in Europe, North America and Australia, but with the introduction of Green Revolution wheat and rice dwarf varieties in the 1970s, fertilizers became vital. In 1961, more than 34 million tonnes of synthetic fertilizer was applied, a balance of nitrogen, phosphorus and potassium (the familiar N-P-K mix). In 2014, almost 280 million tonnes of commercial fertilizers was applied, with a tenfold increase in nitrogen and about fivefold increases in phosphorous and potassium. More than half of the fertilizer is applied in Asia; fertilizer application rates have actually declined significantly in Europe and North America as more precise dosing and delivery has been developed. Africa, and to a lesser extent Latin America, applies too little fertilizer to optimize yields. More than 80% of total fertilizer is used for five crops: maize (40%), soybeans (22%), potatoes (10%), cotton (7%) and wheat (5%).

Different crops and regions use different herbicides and different application rates, so there is no single trend in use. Most of the acreage is used for major field crops in developed countries, including maize, soybean and a range of GM crops, such as Roundup Ready™ soybeans, maize, cotton and canola. Wheat in contrast had fewer chemical options and had lower levels of treatment. Potatoes and vegetables extensively use herbicides.

In most cases in developed countries, farmers simply replaced old chemicals with those herbicides that paired with HT crops. In contrast, in Africa, Asia and South America, farmers seldom used herbicides before GM crops – farmers either weeded by hand or simply accepted the yield penalties from weeds.

More recently the focus has shifted to other uses for chemicals. The introduction of insect resistance, through the delivery of plants that express genes from *Bacillus thuringiensis* (Bt), a naturally occurring insecticide, has worked to lower application rates on maize and cotton, two particularly buggy crops. In those countries that extensively used chemicals to control bugs in cotton, Bt GM varieties enabled farmers to apply far fewer chemicals with little or no change in yields (but higher profits), while in countries that did not previously use chemical applications, yields (and incomes) rose sharply with the use of Bt varieties.

As an aside, we should note that organic farmers are not exempt from the chemical revolution in agriculture. While global organic standards prohibit the use of most synthetic chemicals in organic fields, organic producers still use a wide variety of natural pesticides to control weeds, insects and disease. The list includes a range of natural chemicals, including mulch, dairy cultures, vitamin B, copper sulfate, elemental sulfur, borax, borates, lime sulfur, kaolin clay and various forms of alcohol. A number of chemically derived synthetic options are allowed and used, including spinosad (derived from a soil bacteria), synthetic pyrethrums, chemically derived vitamin C and Bt itself. The bacterium Bt, the same trait engineered into GM cotton, maize and soy, is the most widely used pesticide, accounting for 90% of the organic pest control market.

One thing almost everyone – conventional and organic farmers, environmentalists, chemical companies, regulators and scientists – agrees on is that chemical applications should be used carefully. Public consciousness was raised in 1962 when biologist Rachel Carson wrote *Silent Spring*, which highlighted the risks from chemical use. Her iconic story of the effects of the chemical dichlorodiphenyltrichloroethane (DDT) on the tensile strength of eggs and the role the chemical was playing in endangering the American bald eagle caught the imagination of everyone. As scientists, farmers, industry and government came to grips with her clarion call for more care in the use of chemicals, governments working with scientists constructed the architecture of what we know as the risk analysis framework. Most of us think we have had a set of rules for dealing with risk forever. In fact, the US Environmental Protection Agency was only created in 1970, with the explicit mandate to regulate agricultural and industrial chemicals. The first formal risk assessment was undertaken in the 1970s and it wasn't until the 1980s that the US National Academy of Sciences explored and codified their processes of risk assessment, risk management and risk communications, in the Red, Blue and Orange Books respectively, known by insiders by the colour of their covers. Over the succeeding decades, the norms laid out in the USA transferred to other nation states and were encoded in the Codex Alimentarius, which underpins international norms that are used to assess disputes taken to the WTO and other international agencies.

## The Changing Role of Chemicals in Farming Today

Weeds typically germinate earlier than seeded crops, grow at faster rates than the surrounding crops and produce far more seeds. Weed control is, hence, necessary for the cultivation of crops. Weeds like palmer amaranth, a particularly noxious weed in the southern USA, can produce up to 1 million seeds per plant per summer while kochia, a common weed in Western Canada, is capable of producing 25,000 seeds a summer. A good-yielding variety of wheat typically produces 25–30 kernels of wheat per plant per season, with an exceptional yield of 40 kernels. When weeds are able to produce 1000–25,000 times more seeds than a crop, one can see why poor weed-control practices can quickly become a big problem for farmers. If left uncontrolled, in a matter of 2–3 years weeds would make it virtually impossible for a seeded food crop to compete with the thousands of weeds surrounding it.

If not properly controlled, competition from weeds lowers crop yields. For small farmers in developing countries, poor weed control has been identified as the single biggest cause of low yields. Weeds germinate earlier and grow faster than most crops, allowing them to use the soil nutrients and block the sunlight from nourishing the planted crop, meaning reduced crop yields. Yield losses of 20–80% when proper weed control practices are not followed are common in parts of sub-Saharan Africa. In Canadian fields, yield losses are not near the levels in Africa but are higher than one might expect. Wild oats have been a problem weed for decades in western Canada fields and, in a worst-case scenario, have reduced wheat yields by up to 38%. The more typical yield loss due to wild oats ranges from 5–20%. Canadian thistle is another problem weed: a single thistle plant can, over the course of a single summer growing season, grow to cover a circle six meters in diameter. Left uncontrolled, weeds are capable of inflicting substantial economic, not to mention environmental, damage to a farmer's fields.

The potential for weeds to substantially impact yields when left uncontrolled highlights the importance of the proper use of chemicals in today's agriculture. Farmers have widely varying practices for when they will apply chemicals to control weeds in their fields. Many will apply a pre-seed application to kill any weeds that have started growing prior to seeding so that these weeds do not take moisture and soil nutrients away from the seeds as they germinate. Another application is usually made a few weeks after the crop has been seeded, while the plants are small. This application controls the vast majority of the weeds for the remainder of the growing season, with any that germinate and start to grow after this being few enough that they will not have an impact on the overall yield of the crop. A final application of chemical is made by some farmers just prior to harvest to kill any remaining weeds in the field so that they do not go to seed when the crop is harvested.

Weed control has dictated crop rotations for decades. In the Canadian prairies, the standard crop rotations from the 1950s to 1980s involved a crop–crop–summerfallow cycle to manage for low and infrequent moisture. A crop grown on summerfallow would be harvested and the following year a crop would be seeded into the stubble of the first crop. In the third year, the field would be tilled throughout the summer to try to control the weeds. During this period, the chemicals that were sprayed on crops had some ability to control weeds, but the rate of control was not always what farmers desired. I grew up in the southeast corner of Saskatchewan and the rule of thumb for farmers in this area was to till each summerfallow field once every two weeks to control the weeds. Over the course of an average summer, this would result in 10 tillage passes. Each pass emitted greenhouse gases from the tractor and released carbon from the soil, each of which would vary depending on the amount of rainfall received and the soil type. Most of my friends and I grew up spending our summers driving a tractor back and forth across one field of summerfallow or another, tilling the soil to a depth of 3–4 in. to ensure that the weeds were killed before setting their seeds. Some farmers used soil-incorporated chemicals before or during seeding to supplement other strategies.

The efficacy and safety of chemicals has improved as soil-incorporated chemicals have largely been removed in favour of foliar applications. By moving to topical foliar applications, farmers are able to spray without disturbing the soil, which conserves moisture and reduces erosion, while the newer, high-clearance sprayers are considerably more fuel efficient than the large tractors that I used in my youth to pull tillage equipment. Farmers have different options depending on the crops and traits available. We looked closely at the changes in farm practices in western Canada following the introduction and adoption of GM HT canola. A number of factors were at play. The new GM HT varieties were paired with classes of herbicides that cause higher mortality in weeds, had lower toxicity and remained active for shorter periods. Moreover, over this period many farmers adopted new, more sophisticated sprayers with new tillage practices. With the total production acreage roughly equal between 1995 and 2006, we found that the amount of chemicals applied to canola dropped by 38% and the calculated total environmental impact (on famers, farm workers, consumers and the ecology) of the active chemical ingredient applied decreased by 53%. The amount of active ingredient applied to fields to control weeds dropped by an estimated 1.3 million kg over the period. The change in chemical toxicity has a substantial beneficial impact on the ecology.

By now anyone who has been following the debate about GM crops will be puzzled. The prevailing view is that since GM crops have been first commercialized the amount of chemicals used has gone up. Without a doubt, total chemical use in the global agri-food industry has increased, in part due to the adoption of modern improved cultivars in Africa and Asia. We also can see a rise in North America. Many researchers focus on the use of glyphosate, which is the generic name for the chemical tied to Roundup Ready™ crops. Crops that are tolerant to Roundup are capable of being sprayed by this chemical, without harming the crop, while killing the weeds in the field. Looking at that crop alone, Charles Benbrook showed that use has increased every year since 1995, when GM crops were first commercialized. Yet we assert GM technology reduces chemical use. Who is right? How can we reconcile the estimates of declines for canola with the overall rise in chemical application? The missing variable is the number of acres that we actually crop. In western Canada, when farmers relied on tillage as their sole means of controlling weeds, millions of acres of land would be in summerfallow each year. Prairie summerfallow peaked in 1970 at almost 37 million acres. In 1995, when GM canola was first commercialized for production in the prairies, summerfallow acres had dropped substantially, but still amounted to over 16 million acres. By 2012, there were only 1.6 million acres of summerfallow in the western Canada in 2018. The more than 35 million acres removed from summerfallow since 1970 have been directed to a range of crops and permanent land cover. Between 1995 and 2018, a good portion of more than 14 million acres released from summerfallow has gone into canola, almost all of it using GM varieties. As a result, annual acreage of canola has risen by almost 9 million acres annually, or almost 100%. Thus, chemical use is up, not because of the introduction of GM HT crops, but due to the long-standing and positive shift to minimum or zero tillage and continuous cropping.

What we found in western Canada is also being observed in other major agricultural producing areas. A range of studies have assessed the global impacts of chemical use, finding that chemical use (especially when measured in terms of active ingredient) had dropped by something in the range of 35–40% since 1995, with the largest gains in areas where GM crops are adopted. As in Canada, there are offsetting pressures. Some areas are reducing summerfallow in their systems, which is opening up more land for crops while other regions are actually making use of chemicals for the first time, as they adapt and adopt modern improved cultivars to their ecosystems; these changes would, all other things being equal, raise chemical usage, and the rise in glyphosate use follows that pattern. Overall, however, the introduction of insect resistance into seeds has worked to reduce the application of chemicals. Much of the global change is due to the reduction in the number of chemical applications made on cotton fields following the introduction of IR Bt cotton varieties, and to a lesser extent on maize crops using Bt maize varieties. Non-GM cotton fields were typically sprayed 4–12 times in North America and up to 25–30 times per summer in China to ensure control of the insects that attack the cotton bolls; with the introduction of Bt cotton, farmers report they spray on average 1–2 times per season. Adoption of Bt maize similarly largely replaced an array of chemicals used to control for pests in the field.

One of the challenges is that every crop and farming region has different pests, different management opportunities and is at a different stage of development. Therefore, our studies often run the risk of misallocating effects. In an effort to provide a more rigorous assessment of chemical use following the adoption of GM crops, a group of academics from Kansas State, Michigan State, Iowa State and the University of Virginia undertook a detailed assessment of data from the USA for 1998–2011, which goes a long way to control for the diversity of pests, strategies and development. They found that herbicide use (mostly glyphosate) increased in soybean production by an estimated 28% compared with non-GM soybeans. In contrast, chemical use with GM HT maize was only about 1% higher than on non-GM maize crops. Meanwhile, farmers applied 11% less chemical on their GM IR maize than on conventional maize crops. These researchers then went on to examine the environmental impact of the chemicals applied. Different chemicals have different toxicity and have different half-lives in the environment. Using the evidence from epidemiological and risk-analysis studies, they assessed the environmental impact (on farmers, farm workers, consumers and the ecosystem) of GM and conventional crops. For soybeans, they found that even though farmers applied 28% more chemicals to each field using GM seeds, the environmental impact was indistinguishable from fields planted with conventional seeds. In effect, the lower toxicity and shorter half-life of the active ingredient in the environment for the chemicals paired with GM HT seeds fully offset the rising use of GM seeds. Meanwhile, the combined effect of using HT and IR traits in GM maize seed lowered the environmental impact of those crops 10% below crops using conventional seeds and other chemical strategies.

Our research in Canada involved surveying farmers directly to gather first-hand data about chemical use following the adoption of GM crops. One farmer phoned me while we were doing a survey in 2006 and related his experience with chemical use and GM canola. He told me he had been growing canola on his farm since the mid-1980s. At that time, he said the chemicals for controlling weeds in canola were not very good and he had to plant his canola only in clean fields (i.e. ones with very few weeds) as canola has difficulty competing with weeds for nutrients and moisture. He reported he was able to plant Roundup Ready™ canola in his field that had the biggest weed problem and Roundup cleaned up all the weeds. He found his field was so free of weeds that next spring he did not have to spray for weeds at all before planting a pulse crop. He found he only used chemicals on that field just prior to harvest to kill the few weeds that were growing at that point. This was a dramatic change in chemical use. For the first time, farmers were getting such good weed control from GM crops that they were using less chemical in two years than they had used in one year previously. Economically, we estimated that the combination of follow-on weed control and greater ability to rotate crops in fields generated a modest but sustained return to producers over multiple years.

Most farmers are responsible stewards of the land and have concerns about the use of chemicals and the potential for the development of herbicide resistance within weeds. As with any chemical strategy for handling pests – be it weeds, insects or viruses in plants, animals or humans – selective pressure of a single strategy can lead to resistance or immunity in the target pest. In essence, chemicals seldom kill all of the target pest, and those that survive are, all other things being equal, more than likely to exhibit some resistance to the chemical used. We have to be worried about that in all chemical strategies. In fact, the introduction of GM Bt maize and cotton was managed to explicitly address this risk: farmers planting IR crops had to manage refugia (i.e. land planted to non-Bt or non-maize crops) where target insects could live without exposure to Bt, so that there would remain a vital target population that was susceptible to the Bt. This problem is not unique to GM crops but is perhaps exacerbated because of the focus on a narrower range of herbicides and pesticides that works with the commercial seed stock. Herbicide resistance in weeds was first detected in the late 1950s and by 2017, just over 60 different weeds had developed some level of resistance. Over-reliance on a single chemical has the potential to allow weeds to begin to develop resistance. A particular concern that developed following the commercialization of GM crops was the possibility that farmers would rotate HT varieties of maize and soybeans and spray the same chemical year after year in the same field, creating just the selective pressure scientists are concerned about. Some anecdotal reports suggest that this may be happening in some areas.

The agriculture and chemical industries are more engaged now and have increased their farmer communication and education campaigns to explain the importance of proper crop and chemical rotations. In earlier times, the rapid innovations in chemicals (especially during the 1970s and 1980s) led many farmers to think that if weeds became resistant to one chemical, a new one would come along and solve the problem. In the USA, the agrochemical and seeds industries and the USDA have invested millions of dollars to conduct surveys to better understand how farmers alternate their use of chemicals and to provide communication and outreach strategies regarding proper practices for chemical use. The key message in all of this is that there has not been any significant innovation in mode of action in the chemical industry for a long time and none is expected over the next decade, so chemical stewardship is vital to ensuring that herbicide resistance in weeds is held in check.

A farmer's most valuable asset, which is transferred from generation to generation, is the soil on their farm. Farmers are very conscious about the health of their soil and strive to ensure that they are utilizing crop production practices that best protect their endowment. For millions of farmers worldwide, this means adopting GM varieties that reduce their use of chemicals. Other producers may use GM varieties that increase their use of chemical pest control, but these for the most part work with, and make more profitable, conservation tillage practices that for many producers allows for less tillage, improves soil conditions, leads to less soil erosion, and makes crop production more sustainable. Some argue that the relatively robust harvest in the Canadian prairies in 2018, in spite of poor moisture reserves in the spring and low in-season rainfall, was a dividend from the combined package of improved seeds, chemicals and tillage generally adopted by farmers in the region over the last decade.

## Chemical Residues and Food

In the past few years, there has been a rise in the concerns about the presence of chemical residues on food products. Fortunately, government regulators have been very active regarding chemical residues on food products and have established thresholds for safe levels of consumption. Although for most of us the idea of consuming trace amounts of a chemical is unappealing, the food we consume is never delivered to us in a form that is 100% pure. Frequently, there are particles of soil on food products, while blades of grass or other plant matter and even dead insect or animal parts are also common. Certainly, some dirt or grass is more appealing than the thought of a dead insect being in contact with food that we will consume, but our modern grocery stores, and the supply chains that underpin them, do an incredible job of ensuring that the produce that is available for sale is as free of these items as possible. Thus, as consumers, we tend to become complacent and forget that these items can, and will, be present.

Most of the chemicals used in field have relatively short efficacy and break down into their elemental components before harvest. The new chemicals are better than the older ones. Chemicals used late in the crop season can be, and sometimes are, detected in the food system. Farmers use these chemicals for two reasons: some crops are sprayed late in the summer or just before the harvest to kill any weeds in the field to control for the next season, while other farmers may apply chemicals to artificially ripen immature crops, a practice known as desiccation. The term desiccation means to 'dry up'. In crop production, desiccants are used to dry the plants to make them easier and more uniform to harvest. The optimal window for harvest is very narrow everywhere, but especially in northern climates such as Canada. Farmers need to get their crops off as quickly as possible to ensure they get the highest quality and volumes. When a crop is harvested using a combine, the cut section of the plant is thrashed in the machinery, removing the seed from the straw. If the straw is too wet (as would be the case with immature crops), it becomes difficult to properly separate the grain from the chaff, so that less grain is harvested as it passes through the combine harvester.

There are two types of desiccants: true desiccants and pre-harvest herbicides. Both are applied to a mature plant prior to harvesting to achieve relatively similar results. In addition to physically drying the plant, desiccation reduces the risk of both weather damage (i.e. frost) and 'shattering' loss when a crop prematurely disperses its seeds. A true desiccant is a contact herbicide which rapidly kills and dries the above-ground portion of the plant over a few days. A pre-harvest herbicide is a slower-acting desiccant which is absorbed and translocated to the root and kills the plant from the root up, which can take up to three weeks.

Should we worry about chemical residues in our food? Many of us do. Mothers of newborns, parents of children with allergies and those of us with other health conditions are especially concerned about what we eat. Canadians have less need to worry than many others. This is because Health Canada aggressively regulates the maximum residue limit (MRL) of pesticides that are permitted on our foods. Scientists in and out of government run tests to determine whether a chemical could be toxic and cause harm or death. Sometimes theory tells us what to look for, but we always try out our suspicions on representative animal populations (often mice and rats, because they share much of our DNA and digestive system), using a dose–response model. We will not go into the full science behind this method but suffice it to say that thousands of public scientists and regulators work tirelessly to refine and sustain that approach. Experiments are then designed to deliver a dose of a suspected toxin to test animals to generate the response: harm such as organ damage, reproductive difficulty, nervous system impairment or death. Once they determine toxicity (but not all substances cause harms), they will then work to define lethal dose, at which at least 50% of the test animals die. The no-observable-effect level (NOEL) is also identified. Once established, a regulator would often then impose a safety factor, usually reducing the maximum acceptable level of any toxin below the NOEL, often to a trace amount. The acceptable levels may be defined as the NOEL divided by some multiple, either 100, 1000 or more.

In Canada, and most other Organization for Economic Co-operation and Development (OECD) countries, the MRL will be set well below any amount which could be expected to cause health issues. Generally, a 0% tolerance for chemical residues is not feasible. Virtually every food we eat has some level of natural or synthetic chemical residue on it, even organic foods. Reglone and Roundup, as the most widely used true desiccant pre-harvest herbicides, are most likely to be found in foods. Both chemicals are currently listed by regulators as acutely toxic in high dose, but both are rapidly eliminated through our digestive tract. Canada currently allows a maximum of only 0.2 mg of diquat (the active ingredient in Reglone) per kg of dry lentils or wheat, compared with the lethal dose of 120 parts per million for rats. Glyphosate residues are allowed at four or five parts per million for lentils or wheat whereas the toxicity level for rats is over 5600 parts per million. These residue rates are judged by regulators to be 98% below the level of potential harm (NOEL). Putting this into practical terms, assuming humans are as sensitive to chemical residues as rodents, we would need to eat roughly 600 kg (1322 lbs) of lentils or wheat over a very short period of time to raise lethal concerns regarding diquat. Clearly that is impossible. Chemical residues are more common on products that are directly consumable, such as fruits and vegetables. Insecticides are often sprayed on fruits and vegetables right up to the harvest to ensure that insects do not damage the crop. Organic crops are not exempt from this treatment. A 2014 Canadian analysis of data provided by the Canadian Food Inspection Agency revealed that 46% of a sample of organic fruits and vegetables tested positive for pesticide residues, with 2% of the products violating Canadian regulations for allowable pesticide residues. For those of us with diverse diets though, it would be impossible to consume any dose high enough to create any measurable risk. There are far higher risks than the chemicals to worry about.

While the scientific and regulatory communities have mostly reached a consensus on the safe use of agricultural chemicals, there are a number of dissenting voices. Glyphosate is a flash point for many. The general consensus of the peer-reviewed scientific literature is that glyphosate is not toxic because the chemical is not absorbed by our digestive tract; instead one would expect anyone exposed to or ingesting the chemical would excrete it harmlessly in their urine. However, a number of recent controversies have challenged the consensus. In 2014, an organization called Moms Across America announced that they had found evidence of glyphosate in breastmilk. There is limited detail about their methods, but they said they collected breastmilk samples from 10 women, had them tested by a commercial lab and found three of the samples contained detectable (they argued high) levels of glyphosate. In response, a group of researchers, led by Michelle McGuire at Washington State University, took on the role of replicating the Moms Across America study, adhering to the norms of the scientific method described in Chapter 1 (this volume). Their study tested 41 samples of breastmilk and 40 of urine from lactating mothers living in the states of Washington and Idaho, both highly agricultural regions in the US Inland Northwest; 42% of the participants reported that they ate strictly or mainly organic foods. Each sample was tested twice, once by an independent third-party lab and once by Monsanto, the developer of the Roundup Ready™ technology. Both sets of results were identical: there was no detectable glyphosate in any of the 41 breastmilk samples but, as expected, 37 of the 40 urine samples tested positive for glyphosate. In eight samples, the level was so low that advanced testing was unable to quantify the glyphosate concentration. This left McGuire's team 29 urine samples with quantifiable traces of glyphosate; their analysis concluded that the highest level of glyphosate detected in the urine samples was 4500 times lower than the regulated safe level for glyphosate.

More recently the international policy community and the courts have entered the fray. In March 2015, a panel of experts convened by the IARC, classified glyphosate as 'probably carcinogenic to humans (group 2A)'. Other agents classified as 2A include DDT, vinyl fluoride, acrylamide, malathion, indoor emissions from household combustion of wood, emissions from high-temperature frying, consumption of red meat and shiftwork that involves circadian disruption. Clearly, the 'probably carcinogenic' classification is a large tent that involves a wide range of common to rare things, which we are either exposed to or choose to engage with. Following the release of this judgement the industrial and scientific community, and some individuals from national regulatory agencies and policy bodies, challenged both the judgement and the process that led to it, arguing that the committee did not appropriately balance the evidence. Then in 2018, a jury at San Francisco's Superior Court of California returned a judgement in favour of Dewayne Johnson, a 46-year-old groundskeeper for a public school district in a suburb of San Francisco, who had contracted non-Hodgkin lymphoma, a blood cancer that affects the immune system; he had claimed that the Roundup he used caused the cancer, citing the IARC classification of glyphosate as probably carcinogenic. The jury ruled that Monsanto had failed to warn Johnson and other consumers of the cancer risks posed by its weed killer. The jury awarded the plaintiff US$289 million in damages ($39 million for damages and $250 million as a punitive award). Monsanto petitioned the judge who oversaw the trial to set aside the verdict, reduce the award or grant a new trial; the judge denied a request for a new trial but lowered the punitive damages to $39 million. Bayer, which had merged with Monsanto in the interim, has launched an appeal. In financial filings, Bayer disclosed that lawsuits from approximately 9300 plaintiffs were pending.

These recent developments outside the 'normal' science system are hotly contested by those inside the system, who argue that NGOs, ad hoc committees and courts are poor judges of scientific risk. Regardless of whether the scientists are right, they no longer define what many people think about the chemicals we use in agriculture.

## The Future for Chemicals in Farming

There have been calls from various environmental organizations to ban the use of all synthetic chemicals in relation to food production, which would lead to many unintended consequences and actually work against the goals of many of those who are most concerned. To put it simply, it is a naïve approach. Remember hand weeding would consume the efforts of more than half the working-age population in the USA, and the loss of all the other technologies and chemicals used in the global agri-food system would lower yields, reduce quality and increase volatility in markets, which would adversely affect the poorest and most vulnerable in the world. We could go back to a world of lower yields, greater shares of our income going to food and greater food insecurity. We find little that attracts us or would benefit any but a few ideologically committed individuals and groups in a handful of the wealthiest and most secure communities and societies. The majority of people in the world would be worse off.

The great strides we have made in feeding a global population of more than 7.5 billion people are all due to the ingenuity and success of our farming sector. Yes, some of the technologies, products and processes were the result of investments by governments and for-profit firms outside the sector, but their success has depended on farmers using what they develop. Agricultural economists Graham Brookes, Farzad Taheripour and Wallace Tyner estimate a global ban on Roundup, the most common currently used agricultural herbicide, would cost farm income worth US$6.76 billion, the result of 18.6 million fewer tonnes of soybeans, 3.1 million fewer tonnes of maize and 1.4 million fewer tonnes of canola. With the loss of Roundup, farmers would have to use other, less effective, chemicals, resulting in a rise in chemical use by 1.7% or 8.2 million kg of chemical active ingredient, which would ramp up the environmental impact from these chemicals by 12.4%. The increased greenhouse gas emissions from the additional chemical applications and higher carbon emissions from the increased tillage to control weeds would be the equivalent of putting 11.8 million more cars on the roads. Calling for a ban on one chemical will simply result in the increased use of chemicals that have higher environmental impacts and are less effective. Nothing happens in a bubble; for every action, other reactions occur. These might seem a small price for banning Roundup, but right now it is arguably the safest and one of the most effective and valuable chemicals in farmers' arsenal. Based on the evidence, it would be one of the last chemicals to attack, not the first up against the wall.

While eNGOs call for bans on existing chemicals in agriculture, industry and science are seeking to provide a more compelling solution to the challenges of farming, through innovation. The combination of artificial intelligence and robotics is revolutionizing chemical use in agriculture. One approach is to improve delivery of chemicals to improve efficiency and reduce any unintended effects. Large, high-clearance sprayers currently used in bulk-commodity fields spray the entire field to control weeds. Companies are now using cameras on spray booms to visually control the nozzles to only spray when the camera 'sees' a weed. A sprayer's computer can be programmed to identify specific weeds so that when the camera detects an unwanted growth, it is identified, and the respective nozzle, loaded with an appropriately matched chemical, is turned on to provide a small, short blast of spray to the individual weed. Some experts speculate that if this technology is adapted to sprayers used to spray bulk-commodity fields, it could reduce chemical use by 80–90% with no loss in weed management efficiency.

Innovations in crop agriculture have resulted in more acres of crop land than was previously the case. These additional acres of crop, over 10 million in our home province alone, are often falsely cited as proof that GM crops increase the use of chemicals. They can only sustain that argument by ignoring that acreage has increased dramatically, or that insect and plant disease ranges have increased due to climate change, neither of which has anything to do with GM crops.

Given the opportunity, farmers would prefer not to have to use chemicals. Currently, however, abandoning chemicals would decrease yields to the point that many producers would no longer be profitable and would cease marginal production, dramatically reducing food security for everyone. Chemicals are an essential part of economically viable and sustainable farming. Technological advances in agriculture have dramatically reduced the amount of chemicals that are used in agriculture, especially to control weeds. As the climate changes, traditional zones of insect and plant disease are changing, making continued innovation essential to ensure that adequate amounts of food are produced. Without innovation to create insect- and disease-resistant plants, higher levels of more toxic insecticides and fungicides will need to be applied to crops to defend against insects or disease.

Chemicals are used in all forms of food production. Even though the organic industry claims that chemicals are not used in the production of organic food products, organic regulations allow for, and organic farmers do use, chemicals. Government regulators rigorously review the safety of chemicals to determine levels for chemical residues on food products that are well below levels of potential harm. While many consumers may prefer the idea of foods produced with zero chemicals, they change their mind when they are offered often stunted, diseased or buggy fruit and vegetables that have not been sprayed. Chemicals are an essential part of the safe production of food products and, without them, yields would decline dramatically, food prices would rise and food insecurity would increase.

## Further Reading

Brookes, G., Taheripour, F. and Tyner, W.E. (2017) The contribution of glyphosate to agriculture and potential impact of restrictions on use at the global level. *GM Crops & Food* 8(4), 216–228. DOI: 10.1080/21645698.2017.1390637.

Klümper, W. and Qaim, M. (2014) A meta-analysis of the impacts of genetically modified crops. *PLoS ONE* 9(11), e111629–7. DOI: 10.1371/journal.pone.0111629.

McGuire, M.K., McGuire, M.A., Price, W.J., Shafii, B., Carrothers, J.M. *et al.* (2016) Glyphosate and aminomethylphosphonic acid are not detectable in human milk. *The American Journal of Clinical Nutrition* 103(5), 1285–1290. DOI: 10.3945/ajcn.115.126854.

Smyth, S.J., Gusta, M., Belcher, K., Phillips, P.W.B. and Castle, D. (2011) Changes in herbicide use after adoption of HR canola in Western Canada. *Weed Technology* 25(03), 492–500. DOI: 10.1614/WT-D-10-00164.1.

# 11 Plant Vitality: Will 'Superweeds' Strangle Our Future?

*Growing up on a farm in the 1970s and 1980s meant that gardening was a measurement of maturity. At first, you were only old enough to help plant potatoes by dropping the cuttings into the hole. One quickly learned to be sure that the 'eye' was pointing up! Next you were old enough to shell peas and husk corn. Soon after this, you could be relied upon to hill potatoes and to weed the garden. Once I reached my teens, I was called on to do a lot of the garden work, in addition to all the other 'jobs' a teenage boy had on a farm, from milking cows to baling and hauling bales.*

*Now you might think, what's the big deal, looking after a small garden? On my family's farm, the garden was actually three separate gardens. There was one just for potatoes that had to have a minimum of 500 hills planted each year. The other two were for vegetables: peas, beans, onions, radishes, squash, beets, carrots, cabbage, broccoli, cauliflower, corn, tomatoes, turnips, pumpkins, cucumbers, lettuce and spinach. Even with all of these, there was still room remaining for my mother to plant flowers like gladioli and sweet peas. Learning what was a plant and what was a weed happened early in life on my farm.*

*Hundreds of hours were spent each summer planting, watering and weeding the gardens and they delivered the food that my family canned, froze and stored to satisfy our needs until the next harvest 10 to 11 months later. In rural Saskatchewan in the 1970s, the roads were not built up like they are today, and it was common to be stuck on the farm for several days at a time because the only road to the local town was blocked with snow. Going to the local grocery store on a daily basis was simply not the option it is for consumers today.*

*One of the first things I learned about weeding our gardens was just how prolific a single weed seed could be. A single carrot or beet seed returned a single carrot or beet. When I planted a potato, a good yield was maybe 8 to 10 new potatoes. Tomatoes were better and corn, in terms of kernels, the best of all. Weeds produced hundreds of seeds. Take a moment the next time you see a dandelion flower that is going to seed – there's close to 100 new seeds there. Each weed in our garden robbed the vegetables of nutrients and especially water, which in the early 1980s was not plentiful in Saskatchewan.*

*Missing a single weed in a garden could result in dozens, if not hundreds or thousands of new weeds the following year. Miss weeding a couple of these and you will certainly increase your weeding efforts the follow year. The thing that I noticed growing up is that weeds are far more prolific at reproducing than any of the domesticated plants that my family relied upon to feed ourselves throughout the year. As a youth, I became firmly convinced that weeds were super pesky as they simply never, ever went away. No matter how many I pulled one summer, there was a new crop, just as plentiful as the previous, the very next year, if not later the same season.*

## The Need to Control Weeds

It is not possible to put an adjective in front of the word weed that is meaningful. A weed is a weed, plain and simple. A weed is a plant that grows where we don't want it: they grow in all kinds of climates and soil conditions in a way that reduces crop and food production for those farming the land. In some instances, researchers in developing countries have found that weeds are responsible for yield losses of up to 80%. Poor weed control is often cited as the single biggest contributor to low maize yields for African smallholder farmers. The number of weedings required following planting is the main limiting factor to farm sizes in Africa. This is because weeding on the vast majority of small African farms is done by hand or with a hoe, by women and children, just as it has been for thousands and thousands of years – and still is for each of our family gardens. Throughout history, weeds have robbed crops of resources and reduced yields, contributing to crop failures, famines and starvation. Since the domestication of plants by our distant ancestors, weed control has been an integral and necessary part of crop production.

Efforts to suppress and control weeds have been around for millennia. The most rudimentary form of weed control was through selective grazing by certain types of livestock. Cattle, goats and sheep all have lengthy histories of use as a means of controlling weeds, which has met with moderate success. The reality is that there are some species of weeds that no type of livestock will eat and these weeds can only be controlled via human effort. Also, having animals walk through crops causes considerable damage due to trampling of the plants. The use of selected livestock to control weeds is more common in pasture land, where weeds can restrict and kill off the grass intended for other livestock. Repeatedly throughout history, weeds would simply come to dominate a patch of farmable land, resulting in such meagre crop yields that the farmers would simply abandon the patch of land and move to start farming land where these weeds were not so prolific.

By the early to middle of the 19th century, the means of controlling weeds shifted from physical hoeing to a new and innovative means: using chemicals to control weeds. As the Industrial Revolution provided new ways of manufacturing and transporting products, it also provided innovative ways of examining how food production could be improved. Though the ability to ship commodities such as grains, pulses and oilseeds from continent to continent helped to mitigate the dramatic effects of a crop failure, most countries preferred to identify local solutions that could be relied upon as well. To this end, chemical companies were becoming more commonplace. Through trial and error, it was discovered that some chemicals had the ability to kill plants. Admittedly, this was more likely to have been discovered through error from spilling chemicals on the ground at the manufacturing plant than through the process of a controlled experiment (see Chapter 10, this volume, for a full discussion of the chemistry of farming).

At this time, fields were still small, worked by draught animals, hence weeds could be controlled by a person walking through the field and individually spraying them. Nevertheless, weed control was an issue of concern in Europe. By the 1850s, it is estimated that there were several dozen chemicals available for use in Europe. In contrast, farm land was still being broken (ploughed for the first time) in much of North America and weeds were not yet an issue. In a broad sense, little changed for the next 100 years: while farm and field size increased somewhat, the vast majority of the work was still predominantly done through the use of draught animals and humans. However, from the First World War onward, motorized vehicles and implements became available, and the mechanization of agriculture happened very rapidly. By the late 1950s, most farms in the leading industrial countries had transitioned away from horse- and ox-drawn implements to using tractors to pull the same implements.

From then on, the combination of mechanization and a broader suite of chemical weed-control options resulted in larger scale applications of chemicals to fields. As field size increased with mechanization, chemicals became essential to controlling weeds in larger and larger fields. While hoeing weeds by hand was a feasible option 150 years ago when the size of the field was half an acre (equivalent to half of an American football field), it no longer is when the field is over 300 acres in size and many farm enterprises manage multiple fields.

Between the early 1970s and mid-1990s, the common belief was that single modes of action were the best strategy for controlling weeds. What this essentially means is that farmers should use one chemical to control the weeds in their fields. This was partly agronomic and partly business related. The agronomic side of the strategy was that one chemical applied at an efficacious rate was the best way to control weeds. For example, if a farmer needed to control grassy weeds, then a chemical designed to control grassy weeds should be sprayed. Crops at this time were sensitive to chemicals and applications of chemicals certainly could stunt the growth of a crop for a period of two to three weeks – crop plants would turn a slight shade of yellow as they struggled to overcome the toxic effects of the chemical application. Due to this sensitivity, farmers could not afford to take the risk of spraying a crop twice in the same growing season. It also meant that the use of more than one chemical in a tank for mixed application was not very common. Hence, if the farmer sprayed for grassy weeds, they would not be able to spray for broadleaf weeds the same year. The only real option for controlling broadleaf weeds was to use a broadleaf spray the following year, or to summerfallow the field and use tillage to control the weeds. There were some chemicals that could be tank-mixed together that did provide a degree of control of both types of weeds, however they were expensive, and the weed control was not always optimal. One strategy to control broadleaf and grassy weeds in the same season was to spray for one prior to planting and the other after emergence.

The business side of the strategy involved the patenting of the chemical active ingredients and their application. This meant that there was virtually no financial incentive for chemical companies to combine modes of action and control grassy weeds and broadleaf weeds in a single chemical application. The industrial structure added complexity, as one chemical company would have a good chemical for controlling grassy weeds, while another company would have the best one for controlling broadleaf weeds.

As agriculture moved into the 21st century, we saw changes in the approach to farm chemicals. Patents on single mode of action chemicals began to expire, allowing other chemical companies to develop their own generic versions of another company's successful chemicals. At the same time, the practice of mixing chemicals began to be accepted as an efficient means of controlling weeds without increasing herbicide resistance. While weed scientists had been advocating for the mixing of chemical modes of action to control a broader spectrum of weeds for decades, this message took a considerable length of time to be adopted into common practice. As this strategy for weed control became accepted and adopted, the rate of weeds developing herbicide resistance slowed after 2005.

The adoption of GM crops has changed the way we use chemicals in the production of our food. Prior to GM crops, chemicals were sprayed on crops, which could stunt the crop for a period of time. Alternatively, some chemicals were applied directly to the soil through tillage at the start of the crop season, prior to planting the crop. The practice was designed to kill weeds as they emerged from the soil. The challenge of this means of weed control was that some of these chemicals had high levels of toxicity and could have effects on the planting of a crop the following year due to the presence of chemical residues, limiting the crop options available to a farmer. By moving to chemicals that could be applied to the leaves of the plants, agriculture was able to move away from the use of soil-incorporated chemicals that had a higher level of impact on the environment than chemicals that were more benign.

## Herbicide-Resistant Weeds

As farmers incorporated chemicals into the production of their crops, weeds began to develop varying degrees of resistance to the chemicals being used. In the 1970s on our farm as a youth, we engaged in spot spraying around the farm buildings for wild mustard. I remember my father telling me that in the 1950s essentially all you had to do to kill wild mustard was walk past it holding a pail of chemical while 20 years later it required a concentrated application of chemical to kill it. While this is far from a science-based assessment of the development of herbicide resistance in weeds, it affirms that weeds began to develop resistance to chemicals as the use of chemicals to control weeds in crops became a routine part of crop production.

The development of herbicide resistance in weeds is an issue of concern for all those involved in agriculture. The development of herbicide-resistant weeds dates back to the 1950s, 40 years prior to the commercialization of GM crops, hence claims by numerous eNGOs that herbicide-resistant weeds started with GM crops are highly misleading. Moreover, there are many herbicide-resistant weeds in Europe, where GM crops are not allowed to be grown in most countries. Portugal and Spain are the only two countries allowing the production of GM crops and they report 5 and 38 herbicide tolerant weeds, respectively. By comparison, non-GM crop producing countries have herbicide-resistant weeds as well: Italy has 30, Germany 33 and France 50. By comparison, the USA is the leading producer of GM crops and its top three states with herbicide-resistant weeds are California with 30, Illinois with 27 and Kansas with 26. Herbicide-resistant weeds in the top three crop producing provinces in Canada are Alberta with 23, Saskatchewan with 19 and Manitoba with 22. Cumulatively, Canada and the USA have 68 and 161 herbicide tolerant weeds respectively. The non-GM adopting countries of France and Germany have a higher herbicide-resistant weed density, with one herbicide-resistant weed per 13,000 $km^2$ and 11,000 $km^2$, respectively, while densities in the USA, Australia and Canada are one weed per 61,000 $km^2$, one per 267,000 $km^2$ and one per 543,000 $km^2$, respectively. Gaining a better understanding of herbicide tolerance in weeds is such an important issue that efforts are underway globally to track the issue and increase awareness of when and where issues of concern are being observed. A website for the global tracking of herbicide-resistant weeds has been established (http://www.weedscience.org/).

There are 25 different groups of chemicals, as classified by the Herbicide-Resistance Action Committee (HRAC) system, the system with the widest use. Australia has its own system and Canada and the USA share a system as well, both of which are consistent with, but slightly different from, the HRAC system. Each classification contains various types of chemicals that control different weeds. For example, one group of chemicals will be those chemicals that control grass and grass-related weeds, while another will be for those chemicals that control broadleaf weeds such as dandelions.

Weeds have developed herbicide resistance due to farmers becoming too reliant on one classification of chemical year after year. Chemicals are very good at killing weeds, but if a chemical is used too frequently on the same crop in a field, there is an increased probability that those weeds that survive the chemical in that field could exhibit resistance to the chemical used. A weed that becomes herbicide resistant is not resistant to all chemicals; rather, weeds typically develop resistance to just one per class of chemical. Recently, agronomists have found some weeds that have developed resistance to multiple groups of chemicals. Between 2 and 16 weeds are resistant to two or more chemicals in the top 10 agricultural producing countries. Multiple resistance is not yet common, but it is still a concern. One of the strategies that chemical companies are recommending is the use of several classes of chemicals mixed together to offer control for any weeds that might have developed resistance to a single chemical group or to multiple groups.

The agriculture industry is very concerned about the development of herbicide resistance in weeds and has initiated efforts over the past few years to gain a sense of the scope of the problem and to identify ways to help all stakeholders in the crop-production industry understand the importance of proper chemical use and rotation. For example, in the USA, the USDA partnered with the Weed Science Society of America to develop communication and outreach information for farmers. While this will not provide a solution to the weeds that have already developed herbicide resistance, it will aid in slowing future development of resistance.

## Just What Is a Superweed?

So now that we know something of weeds, what is a superweed? While anti-GMO eNGOs would like everyone to believe they are commonly found in every field that grows a GM crop, they are not. Fortunately, nothing could be further from the truth. They don't exist. Very similar to the Loch Ness Monster, many searches have been made, but no superweeds have been found! Superweeds are a boogeyman.

The only fields I have personally witnessed that are infested this badly with weeds are organic fields – some have been so polluted with weeds that I have had to get out of my truck, climb over the fence and walk into the field to be able to determine what crop is growing there. Drive by any field of canola, maize or soybeans in North America and there is a very strong likelihood that it is a GM variety – look for yourself, is it chock full of weeds? I've driven past hundreds of such fields each summer and have yet to see a field that is weedy.

Although the term superweed has no definition or recognition in the agricultural sector, it is a term that the anti-GMO eNGO community has tried to establish in the public consciousness as another concern the public should have about the production of GM crops. Despite there being no evidence linking the commercialization of GM crops to changes in herbicide tolerance in weeds, anti-GMO eNGOs have not slowed their advocacy to remove chemicals from agriculture. There are three specific risks advocates present as concerns. The first is that the selective pressure of chemicals will over time force existing weeds in crops to become herbicide resistant. As discussed above, that is not a new risk nor is it necessarily amplified by GM traits, and the industry and farmers are actively engaged in managing that problem, partly with chemical mixes, partly by applying herbicides when weeds are small and before flowering, partly with crop rotation strategies and partly with greater surveillance and management. Second, some critics suggest that a few traits, such as IR, could cross into weedy relatives of cultivated plants, effectively destabilizing natural ecosystems. Fortunately, there are few weedy relatives of cultivated plants, at least partly because most of these species have been moved and are now cultivated well away from their centres of origin. GM canola, soybeans and cotton are for the most part not cultivated in centres of origin or even centres of diversity. The biology of the GM crop production helps, as the thicker canopy in GM crops reduces the ability of weeds to get the necessary sunlight to survive, so there will be very few weeds in a field for the pollen from the crop to possibly pollinate. Chapter 9 (this volume) discusses the high-profile debate about the apparent transfer of some GM traits from GM maize to teosinte in Mexico; suffice it to report here, while it happened, the traits transferred did not provide any competitive advantage to the native plant, and within a few years had been selected out of the genome by the plants themselves. Third, some critics suggest that GM crops could escape from their fields and become weeds themselves, in effect polluting the natural ecosystem. The EU actually undertook a study of that possibility. They deliberately released a range of GM plants (maize, canola, soybeans and potatoes, among others) into a contained but natural ecosystem in the UK and then monitored them for 10 years. By the end of the study period, these deliberately GM plants (even the perennial tuber plants) had selected out the GM traits, because outside the artificial world of cultivated fields, those traits conferred no competitive edge. While in theory herbicide tolerance and insect and viral resistance might offer some value outside a managed field, those traits came at a cost – commercial cultivars are often less hardy as they have been designed to be pampered by farmers.

The evidence following 25 years of GM crop production – with more than 476 GM events from 26 GM crops (excluding carnation, rose and petunia) involved in 800 approved environmental releases in more than 30 markets – is that pollination of weedy relatives from a GM crop is exceedingly rare. There has been one detection of a GM canola variety (in Canada) pollinating a weedy relative known as field mustard. This was detected, controlled with other chemicals and did not pose any sustained problem to the environment due to alternative chemical control options. Rigorous studies of this potential in GM maize, cotton and soybeans have not found any evidence of this pollination occurring. There is some literature that suggests this may be something that could exist with GM rice varieties, but no laboratory or field evidence that it will happen. The absence of any significant examples of cross-pollination to weedy relatives suggests we got something right in our risk assessment and commercial management of GM crops to date.

This relative success in controlling for gene flow was explicitly part of the roll-out of GM technologies. Gene flow is one of the factors considered in the risk assessment for GM crops. Countries that commercialize GM crops use the OECD risk analysis framework to ensure that the commercialization of the GM crop variety will pose risks no greater than those posed by the commercialization of any other plant variety. In Canada, for example, this is done through the federal Seeds Act that regulates the seed quality attributes of uniformity, stability and uniqueness. The Seeds Act also establishes thresholds for environmental risk aspects such as: the potential of the plant to become a weed or to be invasive of natural habitats; the potential for gene flow to wild relatives; the potential for a plant to become a plant pest; the potential impact of a plant or its gene products on non-target species; and the potential impact on biodiversity. Canada requires each proponent of a new GM variety to undertake research that examines how that particular variety interacts with other related weeds, or how it could become a weed itself if seeds are left in the ground until the following crop year (remember weeds are plants that grow where they are not wanted). Canola offers a good example of the challenge. Canola grows in small, thin pods that shatter quite easily when harvested. As a result of this, seeds will be present in a field the following spring and will grow and compete with any crop seeded there that year. In fact, volunteer canola was listed as one of the top 10 weeds that Canadian farmers had to control in the 1980s, prior to the commercialization of GM canola, and volunteer canola still requires control in fields today. While most canola grown today in Canada is HT GM canola, there are actually GM seeds that are tolerant to one of three different chemical groups. No single canola is tolerant to all three chemical groups, so farmers can simply rotate among the Roundup Ready™, Liberty Link™ and Pursuit options to control for volunteer canola in their fields.

Biotechnology seed firms do not have to repeat this research if they introduce new varieties with the same GM trait; they can cite previous research publications and data to document their evidence. Those countries with equivalent regulatory structures and similar agronomic and ecological contexts often will share data to streamline review; that means that if a company wanted to commercialize a GM variety of maize or soybeans, they wouldn't have to do the same research on gene flow in both Canada and the USA as they could do the research once and include the data in submissions in both countries. This kind of regulatory harmonization is not restricted to industrial nations: in 2016, the East African community, including Burundi, Kenya, Rwanda, South Sudan, Tanzania and Uganda, agreed to harmonize their biosafety policy framework.

If, on the other hand, a GM crop is being commercialized in a country for the first time, the firm that developed the new GM variety will have to undertake studies on pollen flow from the GM variety to the surrounding environment. They will have to gather data that looks at the new GM plant's potential impact on biodiversity as well as the ability of that plant to cross-pollinate with any weedy relatives. Some of this data can, and will, be supplemented by previously published scientific articles that provide relevant information. For example, a peer-reviewed publication on the ability of a plant to pollinate a specific weedy relative could be included in the submission, if that weedy relative was known to exist in the adoption region. As part of the study on biodiversity, one of the main concerns would be the potential for volunteers to persist long term in the environment and strategies to control any persistent volunteers. Other concerns would be to determine if the chemicals used on the crop have significant impacts on the local ecology, such as impacts on pollinators or butterflies.

## The Regulation of Chemicals

All forms of chemicals are required to receive regulatory approval prior to first use and then are subject to renewal every 10–15 years to ensure that any adverse effects that might develop from long-term use are able to be quantified and any risk assessment to be recalibrated. In Canada, the federal Pest Control Products Act, overseen by Health Canada, regulates the approval and renewal of chemicals with the work undertaken by the Pest Management Review Agency (PMRA). In Canada, a pest can be taken to mean any plant disease, insect, plant or animal that requires some measure of control. Through this, the PMRA regulates insecticides, fungicides, herbicides and rodenticides, assessing the science-based evidence of impact of any new chemical submission to determine that any risks fall within what has been scientifically determined as acceptable. Renewal of existing chemicals, which falls to the PMRA, is conducted every 15 years.

The renewal of existing chemicals considers any information about the presence of weeds that have been observed to have developed some level of resistance to the chemical. Some weeds that developed resistance to older chemicals no longer used are not reviewed and renewed; in those instances, once the supply of the chemical that is on the market is depleted, the chemical is considered to be removed from the market. In other instances, where resistance to an actively marketed herbicide has been quantified, the potential for alternative control options would be an important part of the decision. If alternative chemical modes of action are available to control a herbicide-resistant weed, then this would be judged to provide for effective means of continuing chemical registration.

One recent example was the application to renew registration of the popular herbicide glyphosate. In Canada, scientists had identified six types of weeds as having built some resistance to glyphosate. The PMRA reviewed the evidence and decided to renew the chemical for the statutory 10 years as they determined there was still effective weed control through the use of other chemicals able to control any weeds resistant to glyphosate. The EU system had more difficulty with this file. If they had simply considered the science, as in Canada, they would have been justified in renewing the chemical registration for 10 years as well. The use of glyphosate, however, had become a contentious issue in the EU. Many eNGOs in Europe were actively campaigning for a complete ban on the renewal of glyphosate. Meanwhile, farmers in Europe were very concerned at the prospect of losing access to a safe, efficient and relatively inexpensive chemical, which would then force them into using more expensive and toxic chemicals, which in many instances were less efficient than glyphosate. European farmers, knowing they had to compete with farmers in other countries, were unwilling to give up a safe and efficacious chemical. When it became evident in the summer of 2017 that a 10-year renewal would have been defeated when the decision came to a vote in the review panel populated with national representatives, a compromise had to be reached. A consensus was eventually reached late in 2017 on a five-year renewal, which outraged some but effectively kicked the ball down the road to 2022, when everyone hoped a different set of options and conversations might be possible.

## Looking Ahead at Chemical Use in Food Production

When the regulation of products moves away from the internationally accepted science-based methodologies to ones of political expediency, often based in extreme applications of the precautionary principle, it weakens societies. Decades of effort have been spent working to develop and improve national and global science-based risk assessment processes, ensuring that they are current and taking into consideration all the relevant advances in scientific knowledge and technology. The simple fact that it is now possible to quantify the presence of something at parts per billion or even parts per trillion, does not necessarily make things any safer than when we could only detect parts per million. It is the dose and the exposure that matter when it comes to chemicals. If we are exposed to something potentially harmful, but for only a brief period, then we are not likely to be harmed; our infrequent exposure to medical or dental x-rays is one example most of us can relate to. We might also be exposed to low levels of potentially toxic chemicals over long periods of time without any harm; the use of chloride in many municipal water supplies that helps to prevent contamination of the water is an example (interestingly, chloride is not classified by the IARC, even though many of us consume it daily). Simply having chemicals used in the production of our crops and foods is not necessarily a danger to human health, society, agriculture or the environment; it is simply a modern reality in the efforts to feed a population nearing 8 billion.

The use of chemicals is heavily regulated to ensure that both human and environmental health are not impaired. Many of the chemicals we developed and used in the past, such as DDT, caused significant environmental harm, but our system today undertakes rigorous scientific research and has a deep understanding of chemicals, the environment, biodiversity and the interactions of these factors. There is no reason today that we cannot provide for the safe use of chemicals in the production of crops and foods.

The most important chemical in the farmer's arsenal is the herbicide. Controlling weeds in the production of crops has been one of the most enduring issues for farming, previously requiring hours of drudgery in the hot sun, hoeing weeds or picking them by hand. The physical limits both restrained productivity and shackled untold multitudes to a subsistence existence. The reality is that in many parts of the developing world, this is still how weeds are controlled. Fortunately, technologies have advanced so that in many places these activities have been reduced as safe and efficient alternatives for controlling weeds are adopted.

Chemical use in crops is vital to ensuring global food security. All of the nations that are net food exporters are intensive (but careful) users of the full array of chemicals available for their crops. Chemical companies, farmers and weed scientists are all concerned about herbicide tolerance in weeds, which is why considerable time and money is invested in ensuring that chemicals are used properly. Farmers who rely too heavily on one specific chemical have been responsible for the development of herbicide resistance in some weeds. Providing alternative strategies and outreach about the importance of this issue will benefit all in agriculture. No farmer wants to abandon a field due to the presence of weeds or be faced with having to try to remove weeds by hand. GM crops have been of major assistance to farmers – they have allowed producers to use fewer chemicals more safely and more effectively, along the way improving yields, enhancing farm incomes, providing safer farming and enhancing the environment.

The long-term sustainable solution to food security is research, education and adoption of the best that science and technology can offer – not arbitrary and unwarranted bans of safe and efficacious chemicals.

## Further Reading

Canadian Food Inspection Agency (2004) Directive Dir2000-07: conducting confined research field trials of plants with novel traits in Canada. Available at: http://www.inspection.gc.ca/english/plaveg/bio/dir/dir0007e.shtml (accessed 3 August 2018).

Joubert, A. (2000) Weed control by smallholder farmers in Ciskei, Eastern Cape Province, South Africa. In: Starkey, P. and Simalenga, T. (eds) *Animal Power for Weed Control: A Resource Book of the Animal Traction Network for Eastern and Southern Africa (ATNESA)*. Technical Centre for Agriculture and Rural Cooperation, Wageningen, Netherlands, pp. 214–217. http://www.atnesa.org/weeding/weeding-joubert-ciskei-ZA.pdf

Kniss, A.R. (2018) Genetically engineered herbicide-resistant crops and herbicide-resistant weed evolution in the United States. *Weed Science* 66(02), 260–273. DOI: 10.1017/wsc.2017.70.

Qaim, M. (2016) Plant breeding and agricultural development. In: Qaim, M. (ed.) *Genetically Modified Crops and Agricultural Development*. Palgrave Macmillian, Houndmills, UK, pp. 15–38.

Smith-Spangler, C., Brandeau, M.L., Hunter, G.E., Bavinger, J.C., Pearson, M. *et al.* (2012) Are organic foods safer or healthier than conventional alternatives?: A systematic review. *Annals of Internal Medicine* 157(5), 348–366. DOI: 10.7326/0003-4819-157-5-201209040-00007.

# 12 The Future of Food Security: Where to From Here?

*We have argued among ourselves about all the things in this book. While we disagree about many things, we still have found much that we can agree about. Along the way, we have pondered different ways to more fruitfully structure discussion about those things we have yet to reconcile in our own minds. Yelling at each other from behind the theoretical and social barriers we erect is not a way to find middle ground.*

*Generating, developing and applying science and innovation in ways that benefit humanity at local and global levels has never been more important and challenging for decision makers in the public, private and civil sectors. The idea that science can be put to work in solving pressing public policy challenges is heavily subscribed to by think-tanks, governments, universities and industry. In many ways, our collective history is a story of scientific progress driving socio-economic development.*

*Science and innovation in the 21st century are characterized by at least three dominant trends. First, the science underlying the modern economy is far more diffuse and distributed than in the past. Small teams of motivated individuals were at the core of the waves of technological change over the past 150 years. Those teams could draw from a relatively stable stock of global scientific knowledge and use it to solve discrete problems that constrained our economy and society. Now, science is both global and dynamic. Between the first formalized journal article in 1665 and now, the world has collectively published about 50 million scholarly articles and was adding approximately 1.5 million articles a year to that stock. At this rate of publication, the stock of knowledge will approximately double by 2050. Second, global science has been amplified by global markets, where consumers and suppliers aggressively compete to be the first to use new innovations. Rapid adoption can yield large economic and social gains, but often with attendant greater uncertainty about the risks and impacts and sometimes leading to lock-in to technologies that may not be optimal (the academic literature points to the emergence of the VHS standard for personal video rather than the objectively technologically superior Betamax option as an illustration of this). Third, the market – represented by self-serving consumers, governments buying on behalf of individuals, firms sourcing inputs to construct finished consumer goods and the broader social community – is becoming much more engaged and demanding. Isolated, supply-driven innovations risk missing the mark of what citizens and consumers want or will tolerate.*


S.J. Smyth *et al.* (2019). *GM Agriculture and Food Security: Fears and Facts*
(S.J. Smyth *et al.*)

## The Three Faces of Social Controversy

Industry, scientists and policymakers interested in advancing a whole range of new scientifically advanced technologies and products are frequently confounded by the response they get from consumers, citizens and civil society. Bill Nye the Science Guy, an American entertainer and science advocate, asserts that 'science rules', yet in many instances science appears to be trumped by values, interests and beliefs.

Techno-scientific regulatory and decision systems are increasingly supplemented with a range of new processes populated with a diverse set of participants who often have inconsistent goals. Feasible pathways to use for new technologies must start with a better sense of the cognitive challenges involved in making decisions in this newly motivated world.

As individuals, we accommodate, appropriate and appreciate different things depending on our immediate needs. Despite what the theorists might assert, as individuals we often project inconsistent preferences in different parts of our lives. The scale and scope of those disjointed choices are difficult for scientists, industry and regulators to parse. The result is that we frequently address the wrong message to the wrong audience, fuelling doubts rather than moderating concerns.

As we negotiate our lives, each of us is cognitively challenged across our three realities. Individuals (and by extension, institutions and groups) have been found to operate with serious cognitive limitations. We make use of dual processes to make choices; we characterize them as System 1 thinking which produces fast, intuitive and emotional judgements and System 2 processes, by which we engage in slow, rational, rules-based decision making. When we access System 1 thinking, we can be primed to seek or avoid risk in ways that are both inconsistent and changeable, and sometimes increase our risk to hazards from our choices. In our different realities, we tend to apply different cognitive shortcuts when using System 1 thinking, which can and do lead to biased and inconsistent choices.

In our role as consumers, we are motivated as self-interested individuals to optimize utility, in effect seeking choices where the satisfaction of consumption exceeds foregone opportunities. What we choose depends on personal preferences, individual knowledge and capacity. In most settings, consumers have incomplete information, forcing us to make leaps of logic or to seek signals and advice from others. Marketers and policymakers posit that we can support better consumer choice if we can raise basic knowledge of a product and its antecedents or increase awareness of the frequency, scale and scope of risks and returns of using the product. This so-called information-deficit model asserts that more education and marketing will yield better outcomes. Surveys suggest greater knowledge and awareness can influence preferences, but also generate polarized views with people at both ends of the opinion spectrum outnumbering moderates.

A meta-analysis of 193 studies carried out in 40 countries from 1989 to 2005, testing the aphorism 'to know science is to love it', found a 'shallow but broad' positive relationship between knowledge and support, albeit explaining only about 8% of the measured variance in opinions. Experience of higher education and knowledge of general science had only minor positive influence of opinions on specific technological applications. For biotechnology applications, there was a tiny negative but statistically insignificant relationship between understanding and opinion; local or specific knowledge had only some modest influence on the acceptance of GM foods.

One explanation is that we do not process uncertain and unfamiliar things the same way as we do with familiar choices. As outlined previously in Chapter 1, we amplify or mute our concerns based on our experience and exposure. In doing so, we fall prey to a number of cognitive traps, including: risk framing, where we tolerate more risk to mitigate losses but avoid risk when it offers gains; affect, where we tend to be positively biased to things that trigger positive emotions (think fuzzy kittens); and mental accounting, where we compartmentalize our choices rather than normalizing risk and return across our life. As a result, we often present a muddled set of preferences to the market, willingly tolerating higher risk activities such as extreme sports, tobacco consumption and driving our personal vehicles, at the same time as we push back against objectively safer but more uncertain choices, such as commercial air travel, nuclear energy or consuming GM foods.

We are similarly challenged as citizens in our dealings with the state. Conceptually, the state is contracted by its citizens to generate public goods and protect against social and market failures. Our policy and regulatory systems are designed to be evidence based and non-partisan, using a range of decision rubrics, including the internationally accepted risk analysis framework and cost–benefit analysis. The goal is to avoid type I errors (i.e. approving unacceptably harmful things) while at the same time minimizing type II errors (i.e. rejecting otherwise safe and efficacious things). In effect, we authorize our system to process using System 2 thinking.

The challenge is that the living, thinking, feeling humans who populate our governing system are susceptible to all the cognitive biases we see in consumers, with the addition of group think, over-confidence (aka hubris) and professional confirmation biases. The reliability of experts has been tested in a number of ways. A famous study involved a large-scale, long-term assessment of the actual frequencies of failure in seven subsystems in operating nuclear facilities in the USA (using data gathered by Oak Ridge National Laboratories), compared with the probabilities of failure assigned ahead of time in the regulatory assessment (using the Rasmussen Report). The authors found significant overconfidence among regulatory scientists: five actual risks examined were higher than the 90% upper confidence boundary of expert opinion and two were lower than the 90% lower confidence boundary; none were within the confidence interval (i.e. the range where one is confident that true values are contained within). Other studies have demonstrated that most regulatory processes are far from consistent. The marginal cost of saving a quality-adjusted life year (QALY) across a range of policy interventions varies widely, with no convergence on the value of life that courts often use to settle civil disputes. Instead, regulatory systems tend to undervalue and underinvest in life-saving measures for familiar, voluntary events such as smoking and traffic accidents, but massively overinvest in preventing risks from nuclear power and GMOs. Objectively, we are suboptimally allocating our energies and resources – with the same outlays, we could save more QALYs by reallocating efforts to mitigate familiar risks rather than exotic concerns.

Finally, people are multifaceted social animals. We belong to many groups that shape and motivate our values and beliefs (e.g. families, churches and advocacy coalitions), interests (e.g. unions and industry associations), and norms (e.g. our professions). Membership has its privileges, as it generates a sense of belonging. Social entanglements, however, that engage our hearts and souls can motivate our reasoning in ways that discount objective evidence. Anyone who is a sports fan knows that we lose our objectivity when discussing our team's performance in a match. We similarly exhibit political and social tribalism in the rest of our lives.

## Sorting out Agreement From Disagreement

Many seem to think that discussions about technology should or could be limited to those with interests – in effect by excluding those who are not directly related to the production, use or consumption of any technology; but that is not going to happen. The rise and expansion of social media has provided a platform for those with minimal interests to gain a voice, but also provided a space for those vocally opposed to situations and circumstances to be highly critical. Some segments of society now rely on one form of social media or another for their information about the world, with the result that many of us have joined cliques and networks motivated to advance a particular world-view. Our norms and policy systems are not designed for reconciling such disparate and diffuse perspectives – we largely rely on local, regional or national systems to aggregate our preferences and reconcile differences, yet the debate has gone global.

Professor of Policy Robert Hoppe suggests we can reduce the scope and nature of any debate into two areas of agreement or disagreement. First, there is the question of cause and effect relationships. This can embody scientific knowledge or social-scientific insight into the stimulus–response relationships that underlie any policy issue. Second, because policy is fundamentally a political act, there is a need to have some clarity and consensus on the norms, values and goals of policy before the necessary coalitions can be formed to act.

The level of consensus among policymakers and policy actors across the two areas influences how well-structured a problem may be. A 'structured problem' is usually characterized by involvement of actors who hold similar views regarding values and ethical norms, knowledge of cause and effect relationships and their interpretation. The resulting high degree of consensus leads to processes designed to resolve a problem by finding the solution. Getting to this perfect world of purposeful effort that converges on the right solution is problematic. A range of criteria must be satisfied, including a self-limiting set of considerations, full and complete information and processes that enable learning and reconciliation of differences. Otherwise, there will be no convergence on the solution.

The 'ill-structured problem' lacks definition in some aspect. Unstructured, badly structured or ill-defined problems are generally controversial and ambiguous in nature, with highly permeable, diffuse or variable boundaries, and solutions that are explicitly political, in that there is no set of facts that will solve the conflict. The actors in these problem spaces lack agreement about cause-and-effect relationships, or at least how one might interpret them, and the goals, values and beliefs they wish to use to resolve the issue or that they want reflected in the solution. Uncertainty dominates, both about the status quo and any future states. Planners Horst Rittel and Melvin Webber asserted that, at the extreme, ill-structured problems are wicked, in that they have no definitive formulation: they are unique; they have no stopping rule; their solutions are not right or wrong but rather better or worse; and every solution is a one-shot operation, such that their implementation changes the nature of the problem itself. The inherent circularity and interconnectedness of these types of problem lead many to simply describe them as 'messes'.

In between these two opposing definitions of well-structured and ill-structured problems are moderately-structured problem types. In some cases, we can agree on the values, norms and goals – in other words, the ends – but disagree about the nature or interpretation of the cause-and-effect relationships. These problems are decomposable, in that one can break them down and amplify our understanding through further research or investigation. Once everything that can be known is identified, structured decision systems, such as cost–benefit or risk assessment, can sometimes find a satisfactory solution that may not please everyone. At other times, we agree about the underlying causal relationships but disagree about what we want to see happen. These types of problems lead governments and others to investigate new ways of engaging, in order first to discern the nature and scope of any disagreement and then to see if areas of disagreement can be whittled away, leaving smaller and more clearly delimited areas of profound disagreement. Often there is no single solution that can satisfy everyone – governments and other authorities sometimes simply need to (or choose to) make hard choices that alienate some.

Although we cannot offer much insight into how or where to do more science that will narrow our disagreements about cause-and-effect relationships, we do have some views and experience with how we can work to better define, and possibly narrow, our disputes over values, norms and goals. In many processes, individuals appear to hold conflicting values without attempting to reconcile or order these values in a systematic way. This ambivalence poses a challenge to policy analysis. Decision makers generally want clear, unambiguous signals on what to do. Faced with ambiguous signals, it becomes tempting to provide advice by suppressing ambivalence or by rejecting it as a symptom of confusion. However, suppressing ambivalence can seriously misrepresent both stakeholders' views on an issue and the possibility of policy solutions that command broad assent. Ambivalence towards new technologies is part of a general uneasiness towards those novel situations that can reasonably be assigned to multiple, potentially conflicting categories or frames. The prevailing interpretation that new attitudes to new technologies constitute an 'about face', overlooks, for example, the fact that the stakeholders who began to oppose biofuels development rarely changed their original position that biofuels are, for example, useful in climate change mitigation. Suppressing stakeholders' ambivalence towards biofuels, by either counting them as opponents to development or writing them off as having no consistent position, overlooks the possibility that there is actually more agreement about shared values than seems to be the case. Instead of the classic 'wicked problem' in which disagreement over the nature of the problem prevents any progress towards a solution, stakeholders may actually agree on problem structuring to such an extent that progress towards consensus is more likely than might be supposed. However, if decision makers persist in ignoring ambivalence, they may end up with the classic mismatch of problems and solutions, where the regulatory framework addresses a problem that is clear to the regulators but not to sections of the public. One commonly experienced result of the 'wrong problem' problem is burdensome regulation on an industry that still fails to reassure citizens that their concerns are being addressed.

The key to accurate problem description is to find a balance between the analyst's need to reduce complexity in order to give clear advice, and the recognition of ambivalence and doubt as an inevitable part of a policy domain that contributes to good policy.

## So What Can We Do?

Knowing our individual and collective foibles is just the start. Education and evidence-based advocacy is probably necessary but is insufficient to realize science-based decision making. We need to find a way to address the cognitive drivers of consumer choice. We need to improve the structure and method for incorporating new voices and moderating cognitive bias in public processes, and create new narratives to link science policy to positive public choice frames.

This new context of global innovation requires a more effective policy toolkit. Most policy systems since the end of the Second World War have aspired to evidence-based policymaking, in which objectively based problems are framed through the so-called public choice, with the burden of proof being on the state to identify a problem and act only where their actions would lead to an objectively better socio-economic outcome. This system is designed to be adjudicated by rational decision makers (within the constraints of available information, cognitive capacity and time), in that they have well-ordered and consistent preferences and will make predictable and repeatable choices.

Deductive, reductionist approaches to evidence-based policymaking are no longer adequate. The traditional framing assumes that largely autonomous actors independently make rational choices that deliver impacts that fit a bell-curve (normal) distribution. All our models assume this; the emergence of complexity confounds that. We now can observe a range of lumpy market and policy spaces, where new structures or densely packed subsystems have to make judgements in the face of profound uncertainty. Decisions from these types of systems are subject to significant asymmetries and network effects, with the potential for uneven outcomes (sometimes described as emergent or non-linear). Where these conditions hold, our conventional conception of policy making leads to biased and inconsistent conclusions, which can lead governments to make poor policy choices. These circumstances, along with concurrent concerns regarding the questionable authority of science (including the so-called 'reproducibility crisis' and loss of trust in 'expertise'), necessitate an expanded toolkit to develop and implement policies that are seen as appropriate and acceptable.

Recently there has been a move to characterize and study policy in the context of evidence-informed policy making, where causal stories frame problems subjectively in the context of the social system and decision making involves 'muddling through'.

This challenge to policy analysis is not unique to the science and innovation space. Nevertheless, the science and innovation policy space engages with all aspects of the new policy paradigm, making it a good place to do exploratory and comparative analysis within and beyond the different issues and tools. A new policy research agenda must necessarily address three discrete but interrelated aspects of governance: strategic assessment, public engagement and decision making.

First, we need to work to reframe how we assess and judge prospective and emergent scientific advancements and technological or organizational innovations. Assessment begins well before we know of the applications. Whereas in the past, governments used to be the lead, setting national priorities and providing the lion's share of funding, the space is increasingly occupied by other funding entities. Foundations, new arms-length granting agencies, universities and private firms are testing a set of new structures to accelerate science, technology and innovation. These include conditional grants, networked multidisciplinary projects, research and industrial clusters, public–private partnerships, big science infrastructure, and institutes, centres and open innovation spaces. These systems have proliferated with little evaluation of their effects. There is a need for new diagnostic, design and evaluation tools. There has been significant work already on the impact of different decision criteria (e.g. scientific novelty, commercial viability, economic impact, sunk costs, environmental sustainability, safety, risk, uncertainty and ethics) on both the nature and flow of selection decisions. Theory and evidence suggest that when ill-defined or nebulous criteria are added to decision systems, the structure and rules we use (sometimes called choice architecture), our cognitive biases (such as risk framing and mental accounting) and our mental short cuts (sometimes called decision heuristics) drive choice more than objective evidence (as discussed in Chapter 1, this volume). This is especially true when different people argue from different starting points, which leads them to perceive the same prospects quite differently, or when evidence is agglomerated without consideration of the relative importance of the source and its norms, values and goals.

Given that less than 1% of funded projects deliver technologies or products that find success in the marketplace, research design, selection and evaluation warrant more consideration. We need to determine whether research management decisions – including the decisions by researchers to collaborate and apply for funding, the panels to adjudicate and funders to allocate funds to specific investments – are appropriately structured, tasked and normalized to handle the complexity of this new world. Poorly designed systems are more likely to generate intuitive choices based on weak evidence that leads to anchoring on familiar or immediate opportunities, resulting in greater rejection of unfamiliar innovations due to their higher uncertainty and overall suboptimal activity. A range of new empirical approaches – including institutional and social network analysis, case studies, agent-based models and behavioural experiments – offers ways to test for the impacts of different structures, rules, process and criteria on assessment.

Second, citizens and consumers are no longer content to be the accepting and undemanding buyers of new technologies and their products. People from all walks of life are seeking, sometimes demanding, a greater role in defining the goals and methods of research and innovation. Governments everywhere have taken up the challenge but generally have not found mechanisms that improve the 'fit' of science, technology and innovation into the social space. Governments have constructed a wide range of processes to engage more people in different ways, but there is limited evidence that these efforts have improved public acceptance or government decisions.

Ultimately, engagement, uptake and use of new technologies is an individual choice, but the social context for the decision can at times be critical. Perceptions of costs and benefits, tolerances for risk or uncertainties, and values and interests, in particular, are personal choices that are fundamentally influenced by the communities and social networks in which we live and work. Most current citizen and consumer surveys ask respondents to explain the personal thought process they go through in making a decision about adopting new production technologies or consuming new products. That misses the important role of our social connections. There is considerable evidence that communities, citizens and consumers are less influenced by the objective evidence of personal cost and benefit than by causal stories and the opinions of others (sometimes family members but increasingly remote celebrities promoting their brand through social advocacy). When people are challenged to deal with conflicting opinions, especially about fundamentally uncertain phenomena, they frequently revert to short-cuts and conventions. We need to delve further into the cognitive and social foundations that drive citizen and consumer opinions (especially by region, socio-economic type or psychological attribute) and to examine the systems we use to aggregate and target their preferences.

Third, we need a new toolkit to design, structure and audit decision making itself. Citizens will not be content to put their faith in others – for many, seeing is believing. There are opportunities to populate the research toolkit. We need to know more about the key organizations that frame decision sets – some have identified new, so-called boundary organizations that are having great influence but are not formally accountable, responsible or transparent. We also need to test for how people use our systems – we know that complicated, densely packed, urgent agendas can force fast, intuitive thinking, but we seldom do anything to control for those outside influences.

There is significant work already done describing and critiquing the goals, structures and outputs of regulatory and policy systems, but little that critically models or assesses the internal decision-making structures and processes and their impact on choices. The rising cost and increased time for review for some classes of technology are unambiguously altering investment strategies and priorities. Specific policies and mechanisms have become flashpoints about the benefits and risks of developing disruptive technologies. New crop varieties, energy developments, transportation infrastructure, new mines and new drugs have all faced long, uncertain and costly reviews and significant social conflict. Decisions can be, and are, skewed by ideological positions and political wrangling in legislative processes, by precedent-setting interpretations of rules during administrative and judicial proceedings and by inadequately supported and unnecessarily complex government review processes. The risk analysis framework, in particular, has framed risk as hazard mitigation, which often truncates any consideration of benefits and skews risk dialogues and decisions to harm avoidance. This approach has arguably minimized errors of commission, but at the expense of errors of omission, as safe and efficacious innovations are delayed or erroneously rejected. Understanding how perceptions about risks, benefits and uncertainties are shaped, and how specific decision events in regulatory systems are structured, offers an opportunity to contribute to science-based, evidence-informed policy and regulation. Theory offers a range of new design, structuring and audit tools – we need to translate those to the decision systems themselves.

There is a real opportunity to build and use a set of theories and methods to assess the relative role of ideas, decision architecture and human cognitive capacity in the regulatory and policy decision arenas that make decisions about which science, technologies and innovations we will pursue and use. Understanding the roles of stakeholders, both as individuals and as principles and agents in purposeful organizations (especially those operating at the boundaries of science, policy and regulation), what they view as evidence and how they use evidence to frame problems, assemble options and make choices within decision-making structures, is fundamental to our future, in the global food system and beyond.

## Further Reading


Allum, N., Sturgis, P., Tabourazi, D. and Brunton-Smith, I. (2008) Science knowledge and attitudes across cultures: A meta-analysis. *Public Understanding of Science* 17(1), 35–54. DOI: 10.1177/0963662506070159.

Guston, D.H. (2001) Boundary organizations in environmental policy and science: An introduction. *Science, Technology and Human Values* 26(4), 399–408. DOI: 10.1177/016224390102600401.

Hoppe, R. (2011) *The Governance of Problems: Puzzling, Powering and Participation*. Policy Press, Bristol, UK.

Kahneman, D. (2011) *Thinking, Fast and Slow*. Macmillan, New York.

Phillips, P., Smyth, S. and Kerr, W. (eds) (2006) *Governing Risk in the 21st Century: Lessons from the World of Biotechnology*. Nova Science Publishers, New York.

Rittel, H.W.J. and Webber, M.M. (1973) Dilemmas in a general theory of planning. *Policy Sciences* 4(2), 155–169. DOI: 10.1007/BF01405730.

Rowe, G. and Frewer, L.J. (2000) Public participation methods: A framework for evaluation. *Science, Technology and Human Values* 25(1), 3–29. DOI: 10.1177/016224390002500101.

Simon, H.A. (1973) The structure of ill structured problems. *Artificial Intelligence* 4(3–4), 181–201. DOI: 10.1016/0004-3702(73)90011-8.

Stone, D.A. (1989) Causal stories and the formation of policy agendas. *Political Science Quarterly* 104(2), 281–300. DOI: 10.2307/2151585.

# Index